The Acacia Heist

HOW THE 2,300-YEAR HISTORY OF ACACIA WAS STOLEN IN THE ACACIA WARS

Mohamad Alnoor

Copyright © 2025 by True Acacia
True Acacia™ is a trademark of BioQuantum Solutions, LLC.
All rights reserved.

No part of this publication may be used or reproduced, stored in a retrieval system, or transmitted in any form or by any means (electronic, mechanical, photocopying, recording or otherwise) without prior written permission of the publisher, exept in the case of brief quotations used in critical articles or reviews.

Printed in the United States of America.
ISBN (Paperback) : 979-8-9995916-0-9
First Edition: July 2025

For more information, or to book an event, contact:
TrueAcaciaMovement@Gmail.com
www.TrueAcacia.com

Table of Contents

CHAPTER 1 1
    THE NAME IS ACACIA 1

CHAPTER 2 11
    THE ANCIENT SONT, SHITTIM AND KHADIRA 11

CHAPTER 3 29
    AKAKIA CONSOLIDATED 29

CHAPTER 4 43
    ACACIA UNITED – ACACIA DIVIDED 43

CHAPTER 5 49
    LES PEDLEY SPARKS THE "ACACIA WARS" 49

CHAPTER 6 59
    THE VIENNA COUP 59

CHAPTER 7 68
    THE MELBOURNE CORONATION 68

CHAPTER 8 78
    IBC REASONS FOR ACCEPTING THE 2003 PROPOSAL 78

CHAPTER 9 81
    IBC REASON #1- MINIMIZING DISRUPTIVE NAME CHANGES 81

CHAPTER 10 94
    IBC REASON #2 – BRAND IMPORTANCE 94

CHAPTER 11 102
    OTHER AUSTRALIAN LOBBY ARGUMENTS 102

CHAPTER 12 107
ECONOMIC SIGNIFICANCE COMPARED 107

CHAPTER 13 122
ECOLOGICAL SIGNIFICANCE COMPARED 122

CHAPTER 14 152
FUTURE ECONOMIC AND ECOLOGICAL TRENDS 152

CHAPTER 15 157
HIDDEN IN PLAIN SIGHT – THE MOST OBVIOUS COUNTERARGUMENT 157

CHAPTER 16 160
IF ONLY AUSTRALIANS ATE REAL ACACIA GUM 160

CHAPTER 17 168
THE NEW ACACIA DECEPTION 168

CHAPTER 18 173
THE ACACIA HERO - EULOGY TO LESLIE PEDLEY 173

CHAPTER 19 177
THE POINT OF NO RETURN 177

A PLEA TO AUSTRALIA 183

ACKNOWLEDGMENTS 187

ABOUT THE AUTHOR 188

DISCLAIMER 189

CHAPTER 1

THE NAME IS ACACIA

In the shadowed memory of our species, before the walls of Ur rose, before the first hieroglyph pressed into clay, before science was divorced from spirit, there stood a tree.

Look closely at any Acacia tree, its trunk twisted, its branches armored in thorn, yet crowned in green even through the searing drought. This is not a plant of weakness but of relentless vitality, of never-ending resilience. It thrives where all else withers, and its gum does not flow from damage but from adaptation, from the mastery of survival.

Long before the language of science split from the prayers of priests, before "Acacia" was stripped away to be filed in a botanical ledger or coded in a gene bank, there was awe, fear, and worship at the foot of this tree. In the world's harshest deserts, where barely anything could survive, Acacia thrived. Its trunk was gnarled and unyielding; its branches bristled with thorns that could rend flesh and armor alike. Yet, from the wounds of its bark, a resin bled, a golden gum that caught the sunlight and the imagination of men and women for millennia.

Not a towering redwood, nor the sacred fig revered in Asia, nor the olive tree crowned in the groves of Greece. This tree was thorned and twisted. Modest in stature, and yet hiding within its sap, its golden tears, something so ancient and profound, it would slumber in silence for millennia.

To the casual eye, it was nothing, a mere survivor. But to the initiated, to the ones who listened in the dark, the Acacia was the secret around which the unseen world turned. In the lore of shamans and the encrypted texts of secret brotherhoods, it became more than wood and leaf. It was the vessel of prophecy, the wellspring of sacred healing, the vault that might hold the last revelation before the twilight of human vitality is sealed.

What secrets were hidden in the resin of Acacia? In the oldest stories, the answer is as elusive as it is constant. It was the "tree of incorruptibility", the only wood, so legends say, that would not rot. It was the source of a healing sap, a balm for the body and soul. In the mystery cults of Africa, in the rites of passage among pastoralist tribes, the resin of Acacia was burned as incense, consumed in secret mixtures, buried with the dead, and offered to the spirits of the land. The Native wisemen called it the "Keeper of the Gut", long before they could speak of microbiomes or prebiotics. They believed that in its resin lay the power to cleanse, to renew, to bring the lost back to purity.

This is not just a tree. This is Acacia. Not the rebranded species of bureaucratic botany, but the true Acacia, born of African dust and baptized in the sun. This is the story of a secret stolen, of a name stripped from its rightful bearers, of a truth buried beneath taxonomies, political agendas, and the silence of forgotten scrolls.

Across ancient religions and secret brotherhoods, in esoteric texts and forgotten rites, the Acacia has appeared as a symbol of incorruptibility, divine wisdom, and immortality. And yet, this symbol has never been fully decoded. What has Acacia guarded for so long? Why was it chosen for the Ark of the Covenant? Why does it mark the grave of the revered Free Masonic master Hiram Abiff? Why, in myth, were deities born inside its trunk? Why did secret societies, from the deserts of Africa to the cloisters of Europe, invoke its symbolism? The answer, dear reader, is not merely spiritual. It is biological, and it is timeless. Let us draw back the veil and uncover what the Acacia is hiding in the shadows.

Acacia as the Axis Mundi: Occult Symbolism

The Acacia appears as a cipher in the codices of alchemists, the grimoires of magicians, and the veiled rituals of initiatory cults. Tree of Life, Tree of Death, Tree of the Crossroads; such titles recur in the writings of those who danced with the shadows [1].

- In Exodus, Acacia (shittim wood) is the material of the Ark, the holiest vessel. Its role in ancient rites marked it as a boundary between worlds [2].
- In Freemasonry, the sprig of Acacia is the emblem of incorruptibility, of the truth that cannot be slain, the soul that persists even as flesh rots away [3]. The initiate, buried in ritual, emerges from darkness and declares, "My name is Acacia."
- Osiris the Ancient Egyptian God was mythically associated with the Acacia tree which in some traditions enclosed his body and symbolized rebirth [4].
- Early Christian writers sometimes compared the Ark's rot-resistant Acacia wood to Christ's sinless humanity. Scripture itself connects Acacia wood with several sanctuary items, including the pillars that carried the inner veil (Ex 26:32). Later devotional authors occasionally extended the Acacia symbolism to the Cross or even to liturgical objects [5].
- In many societies, sprigs of Acacia trees are fastened above homestead gates and burned in divination sessions to repel malevolent forces [6].

There are whispers at the margins of mainstream science of alkaloids [7], and psychoactive compounds detected in certain Acacia species. Matching tales of visions seen by indigenous initiates who ritualized and meditated beneath its canopy.

In the Bible, the Ark of the Covenant was crafted of shittim (Acacia) wood. It held the Law, the tablets, the blueprint of divine order [8]. The same structure that held God's word now points us toward a biochemical order lost to modern living. What truth might there be to these tales? Was there ever real power hidden in the Acacia? A power to heal, to open the mind, to bridge the gulf between the material and the spiritual?

In the lore of the alchemists, Acacia gum was prized as a binding agent in magical inks, to write the words that would not fade [9]. The Rosicrucian society devoted to the study of metaphysics attribute the Acacia as an emblem of immortality. To inscribe amulets, to draft grimoires, to summon spirits, Acacia was believed to connect the material and the immaterial, to fix the volatile [10]. Egyptian hand-wands of the New Kingdom were often carved from Acacia because the thorny tree was linked to protective deities [11]. Its association with Moses' rod and Solomon's Temple found rebirth in European magical texts.

The practical was spiritual. The physical, sacred. From ink to medicine, Acacia bound worlds. Material to immaterial, living to divine. Maybe they all predicted a future where the human body is under siege; bombarded by processed food, polluted air, microplastics, heavy metals, and artificial sweeteners. Our microbiomes, once lush and diverse, are now deserts. Our guts are sterile battlegrounds where the old tribes of bacteria are vanishing, causing a myriad of health issues.

Nevertheless, Acacia calls them back. A gentle fermentable fiber, Acacia Gum restores. Rebuilds. It reawakens the inner Eden and guides you into a prebiotic matrix.

The Veir of Science and a cure for the Final Disease

Let us draw the first veil aside. In the modern age, where truth is dissected beneath microscopes and categorized in peer-reviewed journals, Acacia's mysteries begin to reveal themselves anew, but never without echoes of that ancient awe.

Modern science confirms what shamans and herbalists intuited. The resin of the Acacia Seyal and Senegalia, is not mere sap. It is a prebiotic powerhouse, a substance that feeds the friendly bacteria in the human gut and restores metabolic harmony. In an era beset by the collapse of inner ecosystems, dysbiosis, autoimmune disorders, mental disturbances, and chronic inflammation; Acacia's gum offers a bridge back to the Eden of our biology [12].

It nourishes the Bifidobacterium and Lactobacillus that are the foundations of human health. It encourages the production of butyrate, propionate, and acetate, the short-chain fatty acids now understood as the cornerstone of immunity, metabolism, and even cognition [13]. The diseases of modern civilization, particularly diabetes, depression, and autoimmune collapse, all trace their roots to the gut. The solution bleeds, as ever, from the holy tree of the desert [14].

We stand now at the threshold of collapse. Our bodies are falling apart in silence. The cure was always there. In golden sap. In sacred ritual. In grandmother's tea and priest's incense. The Acacia is not merely food. It is a covenant. A link between our ancestors and our cells. Between body and spirit and its truth has been buried. Hidden beneath bureaucracies, renamed by impostors, dismissed by institutions that fear what they cannot patent. But now, it rises.

African Roots: The Forgotten Custodians

By the time writing was born, long before the Saharan camel caravans, the Acacia trade was flowing north from Kush toward Egyptian temples and apothecaries since the dawn of written history. "Gum of the Sont (Acacia)" flowed northward along the Nile [15]. In the market towns, it became currency; in the houses of medicine, it was a cure.

The Acacia tree's ability to regenerate from a stump, sprouting green again after flame or axe, has become proverb. "As the Acacia lives through the dry season, so we live through hardship." Among some people, to cut a mature Acacia without proper ceremony was to invite misfortune or the anger of the ancestors [16]. In the Savanna, Acacia trees were said to house spirits or serve as the meeting point of the living and the dead.

From medicine to magic, the bark was steeped for fever, the resin chewed for digestion, the leaves applied to wounds [17]. In sacred rites, its smoke was said to drive away evil and summon the benevolent spirits [18].

It is the tree of life. Not metaphorically but culturally, economically, and biologically.

As is well known in all other facets of life, history is written by those who conquer and categorize. In the blink of an academic conference, centuries of reverence were stolen with a pen stroke. A name, a lineage, an identity erased in the cold pursuit of power and profit. The true custodians, the peoples of Africa and Asia were made invisible, their traditions recast as irrelevant, and their trees appropriated for distant purposes.

The Great Heist: Not All Stories are Sacred, Some are Stolen

In 2005, at the International Botanical Congress in Vienna, a quiet betrayal unfolded. Through lobbying, procedural technicalities, and power plays, Australian botanists secured a vote that would change everything. A heist executed in plain sight, not with violence but with minority votes. Through a campaign of lobbying and procedural manipulation, Australian interests succeeded in reassigning the name "Acacia" to a group of their own species, most of which had neither the history, the biology, nor the sacred associations of the tree's true African lineages [19].

The name "Acacia", once belonging to the African trees that birthed civilization's gum, was reassigned. No longer would the original Akakia "Thorn Tree" of ancient Egypt and Greece; true bearers of the historical, biblical, cultural, and spiritual lore hold the title. Instead, it passed to a genus of mostly Australian shrubs, unrelated in utility or cultural weight [20].

The name, along with its legacy, was transferred to unrelated trees distanced on opposite ends of the Indian Ocean. Behind the facade of taxonomy, a deeper struggle raged for history, economic power, and cultural identity.

This book is born from that wound. A wound not only botanical, but spiritual and political. The story of the Acacia is the story of how modernity erases the sacred, how institutions conceal ancient truths, and how the voices of indigenous wisdom are silenced beneath the machinery of progress.

This was not taxonomy. It was theft. The world hardly noticed the sting of this historical and cultural appropriation, but the consequences still ripple for those who see beyond.

The Final Revelation

Not all is doomed. There is a legend, whispered in the desert at the edge of the night. There is a torch passed down from temple to synagogue to brotherhood and to the whispering alchemist. From healer to seeker.

You are holding the unsealing of something long buried. Just as the Ark will reappear in the end times. Just as the scrolls will be read. Just as the desert shall bloom with Acacia again, so too will the body be healed by what was hidden.

The true Acacia has waited long enough. When the world is sick, when the gardens are paved and the spirits are silent. When memory is lost and only suffering remains. Then, in the light after the dark, Acacia will bleed again.

The incorruptibility of the Acacia will outlast the thieves, and the stolen knowledge will, at long last, be restored. This book is not only an investigation. It is an invocation. It is time for the Acacia to reveal itself, through you.

References:

1. Ritner, Robert K. The Mechanics of Ancient Egyptian Magical Practice. University of Chicago, 1993, p. 111.
2. Dan, Joseph. Kabbalah: A Very Short Introduction. Oxford University Press, 2006, p. 42.
3. Mackey, A. G. The Symbolism of Freemasonry. 1869.
4. Wilkinson, R. H. The Complete Gods and Goddesses of Ancient Egypt. Thames & Hudson, 2003.
5. Ambrose of Milan. On the Mysteries (De Mysteriis), §7.35. Translated by J. H. Srawley. In St. Ambrose: On the Mysteries and the Treatise on the Sacraments. SPCK, 1919, p. 54.
6. Liengme, Caroline A. "Plants Used by the Tsonga People of Gazankulu." Economic Botany, vol. 37, no. 2, 1983, pp. 113–117.
7. Lynch, Michael J., et al. "Ayahuasca Analogue Toxicity: Four Cases Involving Acacia confusa Bark." Clinical Toxicology, vol. 58, no. 7, 2020, pp. 678–683.
8. The Bible. Exodus 25:10–13.
9. Thompson, Daniel V. The Materials and Techniques of Medieval Painting. Dover, 1956, pp. 80–82.
10. Jennings, Hargrave. The Rosicrucians: Their Rites and Mysteries. Tinsley Brothers, 1870, chap. 23.
11. Ritner, Robert K. The Mechanics of Ancient Egyptian Magical Practice. SAOC 54, University of Chicago, 1993, p. 111.
12. Slavin, J. "Fiber and Prebiotics: Mechanisms and Health Benefits." Nutrients, vol. 5, no. 4, 2013, pp. 1417–1435.
13. Calame, William, et al. "Effect of Dietary Consumption of Gum Arabic on Fecal Microbiota, Short-Chain Fatty Acids and Gastrointestinal Tolerance in Healthy Human Volunteers." British Journal of Nutrition, vol. 102, no. 10, 2009.
14. Valdes, A. M., Walter, J., Segal, E., and Spector, T. D. "Role of the Gut Microbiota in Nutrition and Health." BMJ, vol. 361, 2018, k2179.
15. Hunter, F. R. A History of the Gum Arabic Trade in Sudan: 1750–1950. Cambridge University Press, 2019, pp. 4–5.
16. Ibrahim, Gamal. "Trees and Ritual Space in Nuba Initiations." Sudan Studies, no. 61, 2020, pp. 35–42.
17. Burkill, H. M. The Useful Plants of West Tropical Africa. 2nd ed., Royal Botanic Gardens, Kew, 1985.
18. Liengme, Caroline A. "Plants Used by the Tsonga People of Gazankulu." Economic Botany, vol. 37, no. 2, 1983, pp. 113–117; Ibrahim, Gamal. "Trees and Ritual Space in Nuba Initiations."

Sudan Studies, no. 61, 2020, pp. 35–42.
19. Orchard, Anthony E., and Bruce R. Maslin. "Proposal to Conserve the Name Acacia with a Conserved Type." Taxon, vol. 52, no. 2, 2003, pp. 362–363.
20. Moore, G. "A Case of Mistaken Identity: Acacia, the International Botanical Congress and the Australian Connection." Australian Systematic Botany Society Newsletter, no. 134, 2008, pp. 14–16

CHAPTER 2

THE ANCIENT SONT, SHITTIM AND KHADIRA

How did "Acacia" come to mean what it means today? Who first named it? And whose legacy does that name truly carry? In unraveling the tangled roots of the word "Acacia", we uncover not just a botanical mystery, but a profound struggle over spiritual meaning, cultural memory, linguistic authenticity and scientific legitimacy.

It's possible that the oldest recorded creation story begins around 2345 BC. This ancient creation story does not start on a formless sea, but inside a tree that bleeds medicine. A spell tucked away in a Pyramid text (Utterance 436) lets the secret slip: "Horus who came forth from the Acacia, the youth who first saw daylight" [1].

These hushed liturgies graft a dark myth that creation gestated in a living womb of an Acacia tree. The Priests of Heliopolis guarded that tree. Heliopolitan theology says the creator god Atum emerged from the primeval mound (benben). Later texts place the births of Isis and Osiris "in the Acacia of Heliopolis".

The Acacia term used in the previous ancient texts is a translation of the "Snt" from the Middle Kingdom or "seshe" from the Old Kingdom, which are two terms the Ancient Egyptians called the tree [2]. Two thousand years before the ancient Greek philosophers inscribed the word "Akakia" into the annals of botanical history [3], and 4,300 years before the International Botanical Congress would one day convene to decide which continent held the right to its name [4]. This particular tree was already deeply woven into the spiritual, medical, and architectural tapestries of ancient civilizations. In the sands of Egypt, in the ritual scrolls of Biblical Israel, across the sacred groves of Kush, and throughout the fire-lit rites of Vedic India, this tree; thorned, durable, resinous, was more than flora. It was a bridge between the mortal and the divine.

The African/Asian Acacia trees (what the Acacia Heisters are trying to force the world to now call them as Vachellia, instead of the proper Acacia) [5], were not mere botanical entities to these ancient peoples. As discussed previously, they were vessels of power, wood for altars, balm for the sick, and the symbolic cornerstone of both temple and tomb. In the ancient pre-Hellenistic period, this holy tree was called Sont in Egyptian [2], Shittim in Hebrew, [6], and Khadira in Sanskrit [7].

Then, during the Hellenistic period the Greeks initially called it the "Akantha" tree. By the time of the late classical period, Greek pharmacologists and natural philosophers encountered these trees and their products, and named them "Akakia" [8], a term which would echo across languages and centuries. This naming, like much of the Mediterranean scholarship, was not an invention. It was an acknowledgment. The Greeks did not create the identity of the Acacia; they inherited it.

In this chapter, we retrace that inheritance. From the earliest hieroglyphic mentions to the philosophical writings of Theophrastus, we follow the name "Acacia". To appreciate this journey, one must first return to where the roots of this tree story began in the Nile Valley. More than four thousand years before Swedish botanist Linnaeus was born.

In the parched sunbaked terrain of ancient Egypt, the Acacia tree, referred to as Snt or Sont in hieroglyphic and hieratic script [2], stood out not only for its physical resilience but for its symbolic and utilitarian centrality to early civilization. The tree's thorned branches and golden flowers were more than just natural features; they marked it as a sacred entity, both feared and revered. Egyptian medical, religious, and funerary literature offer compelling insights into how the people of the Nile Valley conceptualized this thorny tree. Its wood was incorruptible, and its gum was healing. Its presence in cosmology, medicine, and rituals suggests it was seen as a metaphysical material, one that could bridge the human and divine [9].

Archaeological records from the Old and Middle Kingdoms (c. 2686–1650 BCE) also testify to state-sponsored expeditions into Nubia, notably the land of Wawat, returning with large quantities of Snt wood.[16] In a translated autobiography, Sixth-Dynasty vizier Weni describes royal caravans from Nubian areas of Wawat, Yam, Irtjet, Medja to provide Snt wood for building barges and tow-boats, confirming that Acacia nilotica and its relatives were among Egypt's most strategically important natural resources [10]. These accounts further demonstrate that by the second millennium BCE, the Acacia tree had become both a sacred symbol and a practical commodity within Egypt and Nubia's complex societies.

The earliest, unambiguous references to Acacia in Egyptian medical texts appear in the Ebers Papyrus (circa 1550 BCE), one of the oldest and most detailed medical scrolls ever discovered. It includes several prescriptions involving the gum of the Snt-tree, often combined with honey, date flour, or oil to treat a range of conditions from gynecological complaints to skin inflammation [11].

Egyptian scribes called the exudate k^{3}m (k^{3}mw), simply "gum", most likely the sap of the Acacia. Medical papyri mention mixing k^{3}m with honey for plasters and eyewashes, temple incense manuals add this gum to fragrance cakes, and modern GC-MS work has identified Acacia gum in mummy wrappings [12]. One such prescription, translated from the hieratic script, recommends "Snt-tree gum to bind, to preserve, to purify" [13], a triplet of functions that merges the pharmacological with the sacred. The Ebers Papyrus of 1550 BC also lists the Acacia-dates-honey pessary and is regularly cited as the world's oldest chemical contraceptive. Modern lab tests reflect this, showing how fermented lactic acid lowers a woman's pH below sperm viability. In short, the Ebers Papyrus treats Sont gum as a multipurpose astringent and binder. Recommended for stopping blood, shrinking swellings, sealing burns, and even preventing pregnancy.

This all confirms that the medicinal reputation of Acacia gum in the Nile Valley is at least 3,500 years old. Modern researchers have identified the Snt of these texts as either Acacia seyal or Acacia nilotica, both of which produce Acacia gum with natural antiseptic and astringent qualities; these species remain endemic to Egypt and Sudan.

Acacia's significance extended far beyond the physician's table. The plant's appearance in religious myth and temple architecture marks it as a totemic species in the Egyptian worldview. In the funerary and cosmological literature, most notably in the "Book of Going Forth by Day" (the "Book of the Dead"), there are repeated references to an archetypal tree identified as the Acacia [14]. The English translated text recounts that the soul of the deceased approaches "the Acacia Tree of the Children", likely referring to the Divine Children of Heliopolis, mythological offspring of the creator god Atum. In the mythos of Heliopolis, the goddess Lusaaset, "the Grandmother of the Gods", was said to dwell within the sacred Acacia tree. It was believed that she gave birth to the first gods from its branches. In this cosmology, the Acacia was no ordinary plant, it was the womb of the cosmos [15]. This tree was not metaphorical and is confirmed by many to be the Acacia Nilotica, the original thorn tree that would later be called Akanth before it evolved into Akakia by the Greeks.

Although ancient Egyptians had no letter for the modern "A", the shift from Snt to Akakia would occur through Greek transliteration, as we will later see. For now, the name Snt firmly links Acacia to divine presence, medicinal function, and symbolic immortality. This dualistic symbolism of death and rebirth would later be inherited by other cultures and eventually syncretized into Abrahamic traditions and esoteric schools such as Gnosticism and Freemasonry [16]. In its Egyptian origin, the Acacia was no abstraction. It was planted intentionally in temple gardens, its branches harvested for ritual use, and its gum incorporated into embalming recipes. Scientific wood analyses have shown that Acacia nilotica (Egyptian thorn) was occasionally used for New-Kingdom artefacts, including coffins and statues, because of its high density and natural pest resistance.

To the south, in the lands of Kush/Nubia, the tree's significance blossomed further into royal ritual. In the Kingdom of Kush (c. 800–300 BCE), which shared a complex cultural entanglement with Egypt, the Acacia tree retained its elevated status. Excavations from the Napatan period tomb of King Anlamani (ca. 620 BCE) in Northern Sudan have revealed four faience foundation bowls, each filled with a symbolic substance such as natron, ochre, sand, and a brown resinous gum (probably, though not conclusively proven, to be Acacia gum), underscoring the tree derived resin's perceived protective power in the Kushite afterlife rites [17]. This ritual use implies that the resin of the Acacia (then growing wild in Upper Nubia and the central Sahel) was not only traded but deeply revered.

Further evidence of Acacia's value appears in trade documents and Greek travel accounts from the 5th century BCE onward. Herodotus, in his Histories (Book II), comments on the construction of Egyptian riverboats called "baris", which were made from what he terms the "thorn-tree of Egypt". Though he uses the Greek word Akantha, scholars agree that this is in reference to the Acacia tree, whose wood was known for its durability and water resistance [18]. A 2020 article in *Archaeology* magazine titled "As Told by Herodotus" discusses the discovery of the Nile barge ("baris") boats and it matches Herodotus's details including the planks of Acacia wood. The find confirms for the first time archaeologically that the "thorny Acacia" referenced by Herodotus referred to actual Acacia wood in ancient Egyptian shipbuilding.

Herodotus' observation not only confirms the continuity of Acacia's use across Egyptian and Kushite domains but also provides the earliest textual Greek encounter with the tree. In Herodotus' phrasing, akanthē, xulon aphtharton en hydati ("a thorn tree whose wood is imperishable in water"), we see the first bridge between the functional properties observed by earlier Nile valley civilizations and the naming conventions that would eventually culminate in Hellenistic science. His use of the word Akantha is particularly important because it does not yet refer to a specific species but to the thorny nature of the wood. The word will later evolve into Akakia under the influence of Theophrastus and Dioscorides, but in Herodotus, we see its descriptive infancy [36].

While the Greek term for the gum itself (kommi) was widely used for tree exudates (including myrrh and frankincense), it was the Acacia gum (clear, viscous, water-soluble) that soon became distinct among them [19]. Later botanists would remark on the fact that when Acacia gum is dissolved in water, it thickens rather than settles [20]. This chemical signature, a unique polysaccharide behavior, allowed Greek pharmacologists to distinguish it from similar resins. However, even before these characteristics were formally studied, the material reality of the gum, traded across the Red Sea and up the Nile, had become embedded in the commercial lexicon of the eastern Mediterranean [21].

Meanwhile, across the Arabian Sea and into the Indian subcontinent, another name was gaining prominence. In the Vedic texts of ancient India, the "Khadira tree" (now identified as Acacia catechu) was revered for its medicinal and ritual value in the Atharva Veda (~1200 BCE).

Thus, by the middle of the first millennium BCE, Acacia trees and their gums were embedded in the liturgical, medical, and commercial structures of Egypt, Kush, India, and the Near East. Though they were called by different names (Sont, Shittim, Khadira) they referred to the same families of the thorn tree: Acacia species growing wild across the savannahs, deserts, river valleys of Africa and western Asia. These were not marginal plants as they stood front and center in the architecture of ritual, medicine, and state power.

In the next sections, we will examine how African, Jewish, and West Asian pre-Hellenic traditions were finally consolidated into a single term: Acacia. Through the botanical observations of Theophrastus, the pharmacological texts of Dioscorides, and the geographic compendia of Greek, Roman, Islamic and Post Renaissance Europeans. We will explore how the African/Asian thorn tree acquired its final classical name, a name that remained for 2,300 years until its unjust reassignment.

The Abrahamic Shittim

More than a Millennia before the Greek Akakia inscription, Jacob stalked the dunes with a shepherd's staff and a smuggled satchel of shittim seeds. Each hole he gouged felt like a conspirator's oath. Plant these now, the Lord whispered. What was he planting? The Torah calls it Shittim, the wood required for the Mishkan (Ex 25:5, 26:15). Modern summaries speak of Jacob's "Acacia trees". He drove seed after seed into the sand until a covert orchard stretched like a dark sigil across the frontier [22].

At Sinai, the same planks became the building material of the Biblical Ark of the Covenant. The Greek translators who retold the saga could not miss the omen and they rendered etzê shittîm as ξύλον ἀκακίας ("Acacia wood") [23], fixing Jacob's covert orchard into every copy of the Septuagint and every church's canon. Jacob had planted more than trees;

he'd buried a time-locked escape plan whose fruits carried Israel from captivity to covenant.

The Hebrew Bible refers to the Acacia tree by the name shittah (singular) and shittim (plural). These terms are used repeatedly in the Book of Exodus, where God commands Moses to build the Tabernacle and its contents entirely from "shittim wood".

- **Exodus 25:10** – "And they shall make an Ark of Shittim (Acacia) wood."
- **Exodus 26:15** – "And thou shalt make boards for the Tabernacle of Shittim (Acacia) wood standing up."
- **Exodus 27:1** – "And thou shalt make an altar of Shittim (Acacia) wood."

From the Ark of the Covenant (which housed the divine tablets) to the Altar of Burnt Offerings (where sacrifices were made), Acacia wood served as the architectural backbone of what was considered sacred. The species most likely referenced here are Acacia seyal and Acacia tortilis, hardy trees abundant in the Sinai and Negev regions. Their wood was chosen for its resistance to rot and insects, symbolizing purity and incorruptibility [24].

Jewish Geographical Legacy - Abel-Shittim and the Valley of Acacias

- The Israelites encamped at a location called Abel-Shittim ("brook of the Acacias") east of the Jordan River before entering Canaan. The prophetic books retain this memory:
- Micah 6:5 – "Remember what Balak king of Moab consulted... from Shittim unto Gilgal."

- Joel 3:18 – "A fountain shall come forth of the house of the Lord... and shall water the Valley of Shittim" [25].

Jerome's Vulgate retained the Hebrew loan-word, reading lignis setim, although the modern Nova Vulgata re-aligns with the Septuagint as lignis "acacia". English and Jewish versions reflect both traditions: the King James Version preserves "shittim wood", while the JPS Tanakh chooses the more direct translation of "Acacia wood" [23].

The consistency in translation over centuries highlights a linguistic stability that ties the Acacia tree not only to physical temple building but to a continuity of sacred language.

The Indian Khadira Legacy

While Acacia wood shaped temples and arks in the Near East, it was transforming fires and healing rituals on the Indian subcontinent. The Sanskrit name Khadira (खदिर) is among the oldest designations for the Acacia tree (specifically Acacia catechu) and is mentioned throughout Vedic and Ayurvedic literature. In the Atharva Veda, dating to around 1200 BCE, the tree is praised in hymns and invoked in spells. One verse extols the protective powers of an amulet made from Khadira wood:

- "That Khadira amulet which Bṛhaspati hath bound—strong, overpowering—may it guard us day by day." – AV XIX 35.1–2 (Whitney tr.) [26].

This was no mere poetic device; Vedic ritual manuals (e.g., Śatapatha Brāhmaṇa III 5,1) list Khadira among the hard woods chosen for samidh (kindling sticks) and for the yūpa sacrifice. Its slow, smokeless burn symbolizing purity and endurance.

The Charaka Samhita and Sushruta Samhita, foundational Ayurvedic texts compiled between the 7th and 4th centuries BCE, list Khadira extensively for treating ulcers, skin diseases, bleeding disorders, and as a gargle for oral infections. The extract known as kattha (catechu) was produced by boiling the heartwood of Acacia catechu, yielding a powerful astringent resin still used today [26].

- "Khadira is cooling, pungent, dry and astringent... It cures blood disorders, skin ailments, and strengthens the gums." – Sushruta Samhita, Sutrasthana

Khadira's astringent bark and dense wood were used in Ayurvedic medicine to treat digestive disorders, skin lesions, and bleeding. The Charaka and Sushruta Samhitas mention decoctions made from wood to purify blood and improve digestion. In the Arthashastra (c. 4th century BCE), Khadira is listed as a primary ingredient in the production of dyes and tannins, demonstrating its commercial value [26].

It should be noted that this was a different Acacia species, but it was part of the same ancient thorn-wood family, illustrating a trans-continental recognition of the genus' medicinal and ritual power.

By the 4th century BCE, Khadira had become an economic commodity. In Kautilya's Arthashastra, a Mauryan-era treatise on statecraft and economics, Khadira extract (kadira-rasa) is listed among the prized dyestuffs and tanning agents. This confirms the transition of the tree from sacred forest entity to regulated trade good over generations [26].

Khadira name evolves to Acacia

William Roxburgh's Plants of the Coast of Coromandel (vol. 2, 1802) is the earliest printed source that explicitly marries the Sanskrit name Khadira with the plant later formalized as Acacia. Roxburgh describes the tree as Mimosa. Four years later Carl Ludwig Willdenow transferred Roxburgh's species to Acacia catechu (L.f.) Willd., carrying the Khadira synonym directly into Latin botanical usage. Together, these two publications establish the first clear, citable link between the Indian name Khadira and the classical genus Acacia in Western botanical literature, cementing its place in global nomenclature.

The Globalist Greeks and Romans Consolidate the Name

The journey of the African/Asian Acacia tree, from a sacred vessel in Egypt to a named genus in Greek pharmacology, culminates in the writings of two Hellenistic figures: Theophrastus and Dioscorides [36]. These men, both inheritors of the intellectual traditions of Hippocrates and Aristotle, helped formalize ancient botanical knowledge. In doing so, the legacy of a tree whose utility and mystique had already shaped thousands of years of ritual, medicine, and myth was preserved. Their work fixed in the Greek language what had long been known in Egyptian, Hebrew, and Vedic thought: that the Acacia was a tree of power and healing, worthy of its own name.

Theophrastus (c. 371–287 BCE), who is widely considered the "father of botany", was a student of Aristotle and author of the foundational texts Historia Plantarum (Enquiry into Plants) and De Causis Plantarum. In Book IX of Historia Plantarum, he gives an account of several thorny trees, one of which he names Akakia (ἀκακία). He describes it in specific geographic and morphological terms: a tree native to Egypt, growing in the Thebaid region, producing pods that were used in tanning and yielding a gum that flowed even without incision. The wood, he notes, is imperishable, and the tree regenerates quickly even after being cut down; a detail that aligns closely with Acacia nilotica, known today for its vigorous coppicing ability.

After Theophrastus's use of the term Akakia, many scholars have suggested another Greek meaning to the word by deriving the Greek prefix a- (not) and kakia (evil), suggesting not just a botanical designation but a symbolic association: "not evil" or "free from wickedness" [28]. This may reflect the ancient belief in the Acacia's purifying properties especially after Akakia products become more well known in the Hellenistic world that incorporated Egyptian embalming practices, Hebrew tabernacle rituals, and Ayurvedic medicine. As Eva Crane notes in her scholarly analysis of plant nomenclature, Theophrastus' Akakia is "is evidently identical with the modern Nile Acacia (Acacia nilotica L.)," as it matches both the geographical range and the physical properties of that species [29].

Following Theophrastus, Pedanius Dioscorides (c. 40–90 CE), a Roman military physician of Greek origin, composed De Materia Medica, a pharmacological compendium that would remain authoritative in Europe and the Near East for over 1,500 years. In Book I, he provides a detailed account of Akakia, describing its medicinal extract made by boiling the pods of the Egyptian thorn tree. Dioscorides lists astringent, anti-inflammatory, and wound-healing properties, and notes that the resulting extract is useful for treating diarrhea, hemorrhages, and ulcers. The gum, he adds, can be used as a base for eye salves and mouthwashes [36].

Dioscorides' identification of Akakia as a preparation derived from the Egyptian thorn tree firmly links the term to Acacia nilotica. The fact that Dioscorides places Akakia among his most potent therapeutic agents confirms its high status in the classical pharmacopoeia. Fast forward a thousand years in the Medieval ages and Latin pharmacologists, speak of "succus acaciae", while Arabic physicians (Avicenna, Ibn al-Baytar) adopt the same drug from "Aqaqiya", ensuring the Acacia extract remained embedded in both Latin and Arabic medical traditions [30].

Pliny the Elder (23 – 79 CE), in his Naturalis Historia, reinforces this connection. In Book XII he writes: "The best gum is that which flows from the Egyptian thorn; it is transparent, odorless, and dissolves in water" [31]. He further notes that this gum (clearly referring to Acacia gum) was used in Roman medicine, painting, and incense-making. The tree he describes matches the African Acacias of the Nile and Sahel, particularly Acacia seyal, known for its high-grade gum production [32]. Pliny uses the Latin term gummi to describe the resin, but, like Dioscorides, he links it to the Egyptian thorn.

By the 1st century CE, the term Akakia had been solidified in Greco-Roman botanical and medical literature as referring to the gum yielding thorn tree of Egypt. This nomenclature was not generic; it was specific, empirical, and consistent with the long-standing recognition of the African Acacia as being a plant of rare, versatile utility. Greek and Roman pharmacologists were not merely applying a descriptive name; they were acknowledging a millennia-old cultural and ecological reality.

By the early Roman period, the technical term Akakia was firmly linked to gum yielding Egyptian thorn trees, and the resin itself moved in bulk along both the Red Sea maritime route (through Berenike) and Nile caravans descending from Nubia via Elephantine to Memphis. Ostraca from Berenike and Elephantine list "gum of Acacia" alongside frankincense and myrrh, while Dioscorides and temple incense recipes show it had become a staple of pharmacies and cult practice throughout the Mediterranean world [33]. By Galen's time (c. 129 – 216 CE), Akakia was firmly embedded in Greco/Roman therapeutics. Galen prescribes it as a powerful styptic and astringent, useful against bleeding, diarrhea, and ulcerations.

This early stabilization of the name Akakia in Greek usage is critical. It demonstrates that the naming process was neither arbitrary nor recent. The term emerged organically from centuries of interaction with African Acacia species that had been venerated, harvested, and exported long before the botanical renaissance of early-modern Europe. Moreover, its later etymological construction (a-kakia, "not-evil") reinforced its symbolic reputation as a purifying and healing tree, resonating with its uses in embalming, medicine, and liturgical construction.

Notably, no other thorn tree; neither Mediterranean hawthorns nor Arabian senna, received the same terminological consolidation. To be clear, the term Akakia was not applied broadly to all spiny flora. It was reserved for the Egyptian Acacia and its gum producing kin, precisely because of their unique properties and widespread civilizational recognition.

Early Christian exegetes quickly transformed the botanical Akakia into a spiritual emblem. Origen of Alexandria, commenting on the Ark of the Covenant, tells his congregation: "Moses was commanded to fashion the Ark from Acacia wood. The very name ἀκακία, formed from the privative a- and κακία ('evil'), teaches us that the vessel of God's presence must be fashioned from what is 'without evil' and, like that wood, incorruptible; for Acacia neither rots nor is eaten by worms" [34]. The Septuagint already calls the Ark's timber ξύλον ἀκακίας ("Acacia wood"), and patristic writers like Origen seize on both the wordplay (a-kakia, "without evil") and the wood's legendary imperishability (praised by Herodotus) to make the Acacia an icon of incorruptibility in Christian art and theology [35]. Henceforth, a spiritual meaning was derived from the Greek word Akakia and implanted into Acacia's heritage for eternity till this day.

This means that the botanical identification was already embedded in the Septuagint, the Greek translation of Hebrew scriptures produced in Alexandria in the 3rd century BCE. It's here where "Shittim wood" was rendered as "ξύλον ἀκακίας.". This confirms that the Greek speaking Jews of Egypt equated the sacred wood of the Tabernacle with the same Egyptian thorn tree later described by Theophrastus and Dioscorides.

By the early Common Era, the name Akakia had been stabilized across at least three domains: in botanical science (through Theophrastus), in pharmacology (through Dioscorides and Galen), and in theology (first through the Septuagint and later through early Christian writing that imbued it with symbolic spiritual significance) [36].

The term was Greek in form, but African in essence. It preserved the memory of the Sont, the Shittim, and the Khadira, collapsing millennia of regional knowledge into a single lexical vessel to be used for the next 2,300 years. This was not the imposition of Western taxonomy on indigenous knowledge. It was its enshrinement.

This chapter has traced the evolution of that name from the Nile Delta to the libraries of Alexandria, from the Torah to the Bible, from papyri to pharmacopoeias. It has followed the word Akakia as it crossed boundaries of language, empire, and faith. This chapter has demonstrated, through every step, that the rightful holders of the name Acacia are the trees of Africa and Asia. The same trees that gave us gum, medicine, and metaphors for immortality.

That is the truth the modern nomenclature must reckon with, and it is the truth this book will continue to defend.

The next chapter will examine how this name entered medieval Arabic medical texts, transitioned into Latin, and eventually passed into the binomial nomenclature of Linnaeus, only to be controversially reassigned in the 21st century. Our present chapter ends here, with the Greek name Akakia securely established as the rightful continuation of the ancient African/Asian Sont, Khadira and Shittim trees.

References:

1. Kurt Sethe, Die Altaegyptischen Pyramidentexte, vol. 1 (Leipzig: J. C. Hinrichs, 1908), Utterance 436, §§ 786–87; Raymond O. Faulkner, The Ancient Egyptian Pyramid Texts (Oxford: Clarendon Press, 1969), 70; James P. Allen, The Ancient Egyptian Pyramid Texts, 2nd ed. (Atlanta: SBL Press, 2015), 164.
2. Said Eissa, "Acacia Tree Sndt in Ancient Egypt," Journal of the General Union of Arab Archaeologists (2019).
3. Theophrastus, Enquiry into Plants, trans. Arthur Hort (London: Heinemann, 1916), bk. 4, ch. 2, § 1.
4. Kevin Thiele, "The Acacia Debate" (paper, XVIII International Botanical Congress, Melbourne, 2011).
5. P. J. H. Hurter and D. J. Mabberley, "Vachellia nilotica (L.) P.J.H. Hurter & Mabb.," Plants of the World Online, Royal Botanic Gardens, Kew]
6. Avicenna (Ibn Sīnā), Al-Qānūn fī l-Ṭibb, Cairo 1877, bk. 2, fen 2, ch. 21, 65; Eng. tr. Laleh Bakhtiar, Natural Pharmaceuticals (Chicago: Great Books of the Islamic World, 2012), 55–57.
7. "Khadira (Acacia catechu)," Wisdom Library: Ayurvedic Dictionary, 2025.
8. Pedanius Dioscorides, De Materia Medica, bk. 4, fol. 243r.
9. A. A. Moustafa et al., "Insight on Acacia nilotica," Journal of Ecology & Natural Resources 8, no. 3 (2024); A. A. Hussein et al., "Traditional Ancient Egyptian Medicine," Journal of Pharmacology (2021).
10. James H. Breasted, Ancient Records of Egypt, vol. 1 (Chicago: University of Chicago Press, 1906), §§ 323–24.
11. "Contraception through the Ages: Acacia and Honey Pessary in the Ebers Papyrus."
12. Stephen Buckley and Richard Evershed, "Organic Residues in Ancient Egyptian Mummies," Journal of Archaeological Science 28 (2001): 1–7.
13. Ebers Papyrus (c. 1550 BCE), prescription for snt gum (binding, preserving, purifying).
14. Book of Coming Forth by Day (Egyptian Book of the Dead), Spell 125, "Acacia Tree of the Children."
15. E. A. Wallis Budge, The Gods of the Egyptians, vol. 2 (London: Methuen, 1904), 195; Jan Assmann, Egyptian Solar Religion in the New Kingdom (London: Routledge, 1995), 116–18; Rundle Clark, Myth and Symbol in Ancient Egypt (London: Thames &

Hudson, 1959), 64–66.
16. Albert G. Mackey, "The Sprig of Acacia," in The Symbolism of Freemasonry (19th-cent. lecture).
17. High Museum of Art, Ancient Nubia: Art of the Twenty-Fifth Dynasty, description of King Anlamani foundation bowls.
18. Herodotus, Histories 2.96; see "Baris (ship)" modern summary.
19. "Gum (kommi)," in Brill Encyclopaedia of Ancient Greek Language; Dioscorides, De Materia Medica I 94.
20. "Acacia Gum: Chemistry, Properties & Food Applications," Current Opinion in Food Science (2024).
21. Dioscorides, De Materia Medica I 128; Galen, De Simplicium Medicamentorum Temperamentis XII 39.
22. Midrash Tanchuma Terumah § 9; Exodus Rabbah 35 1.
23. Septuagint, Exod. 25:10–11 ("ξύλον ἀκακίας"); Brown-Driver-Briggs Hebrew Lexicon, s.v. שִׁטָּה.
24. Avinoam Danin, Desert Plants of Sinai and the Negev (Jerusalem, 1983), 84–87; Michael Zohary, Plants of the Bible (Cambridge, 1982), 100–103.
25. Hebrew Bible: Num 25 1; 33 49; Mic 6 5; Joel 3 18; Robert L. Hubbard, Joel and Amos (Grand Rapids: Eerdmans, 1989), 91.
26. Kauṭilya Arthaśāstra II 17 4; Central Sanskrit Univ., "Kauṭilya's Arthaśāstra on Forestry" (2023); "Khadira—Uses in Ayurveda" 2012; Atharva-Veda XIX 35 (Whitney tr., SBE 42, 1895).
27. Charaka Saṃhitā Sūtrasthāna 27, trans. Shiv Sharma and Bhagwan Dash (Varanasi, 1976), 486–89; Sushruta Saṃhitā Cikitsāsthāna 4, trans. Priya Srikantha Murthy (Varanasi, 1991), 44–46.
28. Robert S. P. Beekes, Etymological Dictionary of Greek (Leiden: Brill, 2010).
29. Eva Crane, The World History of Beekeeping and Honey Hunting (London: Routledge, 1999), 519.
30. Pliny the Elder, Naturalis Historia 24 57 ("acaciae succus ex Aegypto"); Pseudo-Apuleius, Herbarium 66, "De succus acaciae."
31. Pliny the Elder, Naturalis Historia 12 32 § 65, trans. H. Rackham (Loeb 370, 1952), 139.
32. John F. Williams and Gordon O. Phillips, eds., Handbook of Hydrocolloids, 2nd ed. (Cambridge, 2009), chap. 9, § 9.3.1.
33. Adam Łajtar and Jacques van der Vliet, Elephantine Ostraca of the Ptolemaic Period (Warsaw, 2018), nos. 87–89.
34. Origen, Homilies on Exodus, hom. 6, § 4 (GCS 7; PG 12).
35. William Roxburgh, Plants of the Coast of Coromandel, vol. 2

(London, 1802); Carl L. Willdenow, Species Plantarum, 4th ed., vol. 4, pt. 2 (Berlin, 1806).
36. Theophrastus, Enquiry into Plants IV 2 8; Dioscorides, De Materia Medica I 129; Galen, De Simplicium Medicamentorum Facultatibus VIII; Septuagint, Exod. 25 5, 9.

CHAPTER 3

AKAKIA CONSOLIDATED

From the thorns of the Nile to the scrolls of Greece, the Acacia tree (known as "Akakia" to the ancient world) held sacred, medical, and cultural weight that spanned civilizations. This "Akakia" was the Acacia nilotica and Acacia Seyal, both native to the Nile basin and identified by their use and growth region. It is this name, rooted in empirical observation and centuries of cross-cultural reverence, that was later Latinized as Acacia and came to dominate trade lexicons, apothecary manuals, and religious translations for over two thousand years. The fact that this name was stripped from the very species that defined it, by a skewed administrative minority vote of modern taxonomists, marks not a neutral classification decision but a historical erasure.

The legacy of Acacia was not born in a laboratory or committee. It was carved by Egyptian priests, spoken by Hebrew prophets, sung in Vedic hymns, consolidated into Akakia by Greek philosophers, medicinally globalized by Islamic scientists as Aqaqiya and finally latinized into Acacia in the Scientific revolution. That legacy must be defended.

As the Roman empire matured, its appetite for exotic substances from the periphery intensified. Frankincense from Arabia, myrrh from Punt, pepper from India, and Acacia gum from the Nile region were increasingly incorporated into daily Roman life. Acacia Gum, the resin derived from Acacia Seyal/Nilotica and its cousins, was used in everything from medicine and food preparation to ink-making and textile dyeing. The name Akakia became synonymous with medicinal purity and divine incorruptibility. This was not merely a taxonomic curiosity; it was a linguistic fossil of religious and scientific continuity.

As Latin became the lingua franca of the Western Roman Empire, Akakia would eventually be Latinized into Acacia, preserving both pronunciation and spiritual connotation. Interestingly, Akakia would first be Arabized before its Latinization.

As Byzantium receded and the Islamic Empire started to control trade, a new hybrid name arose for the Akakia. In the 9th century, Islamic scholars documented the medicinal uses of a tree they called "Aqaqia", a direct transliteration from the Greek word "Akakia". Greek medical manuscripts were circulated through Byzantium and later sparked the 9th-century translation movement in Baghdad. There, Ḥunayn ibn Ishaq's team rendered Dioscorides into Arabic, transliterating the heading as "Aqaqiya"; the Greek consonants survive almost unchanged [1].

Early formulary writers such as Sabur ibn Sahl (d. 869) and al-Kindi (d. 873) adopted the new Arabic spelling. Ibn al-Baytar's "Kitab Al Jami li-Mufradat Al-Adwiya wal Aghdhiya" records that the bark and leaves of Aqaqiya are "cold-dry, gently astringent" and "a leaf decoction is taken for fevers and diarrhea, while boiled leaves or bark are applied to ulcers, toothaches and inflamed gums". He also cites that the healing power of Aqaqiya also comes from the pods and gummy exudate of the Egyptian Acacia, echoing Dioscorides yet grounding the remedy firmly in the medical markets of Cairo and Damascus [2].

Al-Zahrawi, in the pharmaceutical section of his Kitab al-Tasrif, repeats these uses, calling it a mild astringent that calms internal inflammations, and he repeats the standard uses given by earlier formulary writers for stopping bleeding, easing dysentery and "cooling" hot swellings. Both treatises employ the precise name "Aqaqiya", never a generic tree description. Both Al-Zahrawi's Tasrif and the earlier formulary of Sabur b. Sahl use the fully Arabized spelling "Aqaqiya", and no other generic names [3]. By the 11th century, the term is standard from Andalusia to Khurasan, showing an unbroken linguistic line that began in Hellenistic Greek and continued into classical Arabic, nearly phoneme for phoneme.

During the 11th Century, Avicenna (Ibn Sina) wrote of its astringent and binding properties, confirming its use for intestinal disorders, bleeding, and skin afflictions. Crucially, he described it as coming from Shajar al-Aqaqiya, literally "the Acacia tree", a clear linguistic continuity from Greek [4]. Even more interesting is that later Renaissance Latin translations of Avicenna's books used the Arabized Greek word Aqaqiya and Latinized it into Acaciae, deeply influencing 15th century European medicinal texts and evolving the preserved term "Aqaqiya" into Acaciae.

Among the bustling caravan routes, and the sharp scent of spices, merchants prized the gum for stabilizing syrups, physicians treasured its astringency, and travelers tucked it into leather pouches as a universal first-aid. In the Arabized world of medicine, the African/Asian Acacia remained at the center of practical and theological attention. The lively endorsements of "Aqaqiya" shows how, centuries before modern pharmacopoeias, Acacia gum was already a global health essential, long before any Australian wattle entered the story.

Greek physicians prescribed it as an astringent and coagulant. Roman soldiers used it to treat battlefield wounds. Egyptian priests used it in ritual perfumes. Arabian apothecaries used it as a suspending agent in elixirs. Indian medics used it to balance internal heat and bind decoctions. These uses match only the gum produced by the African/Asian Acacia which contains the water-soluble, complex polysaccharide structure historically called Acacia gum [5].

By contrast, Australian "wattles" (which would later be classified into the Acacia genus) do not produce the same exudates. Many are toxic or produce tannin-rich exudates unsuitable for ingestion. No historical source before the 18th century ever mentions an Australian wattle in any medicinal or cultural role. Their entire integration into the botanical record post-dates the Enlightenment era's exploration of botany [6].

Thus, the term "Acacia gum" is rooted in a botanical, medicinal, confectionary, linguistic, and trade lineage stretching across millennia. It was never just a substance; it was a standard, a proof of identity. To now suggest that the name "Acacia" rightfully belongs to Australian wattles, which have no historical presence in these traditions, no use in ancient medicine, and no evidence of ritual or commercial relevance until the colonial period, is a grave rewriting of history.

Following the consolidation of the term "Akakia" which was later Arabized into "Aqaqiya"; the tree remained central to global healing and trade. While gum was its chief export, the bark of Acacia nilotica (exceptionally rich in catechol tannins), was equally prized. Medieval tanneries in Fustat and Nubia steeped hides in crushed Acacia bark locally called "qarad". Goitein describes Nubian growers rafting the material down-river. Medieval finds mention Acacia pods and bark among Egypt's principal vegetable tannins [7].

Modern phytochemical research has verified that extracts from the pods, bark and leaves of African/Asian Acacias (especially Acacia nilotica) exhibit strong antibacterial, antidiarrheal and antioxidant activity, owing to their high tannin and flavonoid content [8]. Australian wattles, by contrast, possess a markedly different secondary-metabolite profile. Several contain indole alkaloids or fluoroacetate that restrict medicinal use, and published pharmacology seldom reports comparable gastrointestinal or antimicrobial benefits [9].

Medieval Apothecaries and the Standardization of Acacia Gum

The Seljuk-Turkish formulary Tuhfe-i Mubarizi by Hekim Bereket contains an entry "Aqaqiya", identifying it as the gum of a thorny tree native to Egypt and the Ḥabasha. The text rates it "cold and dry, moderately binding" and prescribes it for diarrhea, spitting of blood, mouth ulcers and loose teeth. A later Ottoman compilation, Enmuzecu't-Tibb by the court physician Emir Celebi, copies the same description under "Akakiyya (Gummi arabicum)" [10].

Meanwhile, the material continued to flow north from Africa, across Byzantine and Islamic trade routes, into Europe and Asia. Arabic pharmacology, heavily indebted to Dioscorides, preserved and expanded the knowledge of Akakia. By the late-medieval period (12th–15th c.), the drug sold as succus acacia (the astringent gum obtained largely from Senegalia and Acacia seyal) had become a routine entry in both Islamic and European formularies. The Latin Antidotarium sive Grabadin (a mid-11th-century rendering of the Baghdadi Aqrabadhin attributed to Mesue the Younger) lists "succus acaciae" among the forty ingredients of Theriaca Andromachi Major, prescribing 3ii (about 8 g) of finished electuary. Antidotarium Nicolai, compiled at Salerno c. 1140 and copied throughout Europe, places "Acacia, succus" under the letter A in its table of simples; Mellon (Germany, 1475) carries the same reading. Tacuinum Sanitatis and Latin versions of Ibn Butlan's health-manual retain the simple "succus acacie / gummi arabicum," giving it a "cold-and-dry" rating and a standard dose of one-half drachm for dysentery or hemorrhage. Each text thus treats Acacia gum as a discrete, tradable medicament with measured doses, never merely as a vague botanical [11].

From Granada to Baghdad, Cairo to Constantinople, the source was the same; gum extracted from the thorn trees of Africa and West Asia. No other region's Acacias (certainly not the Australian wattles) entered these medicinal records. The reason is chemical as well as geographical. The African species exude a complex, nontoxic polysaccharide rich gum that is both edible and medicinal; a profile absent in most Australian taxa [12].

The European Renaissance Spurs interest in Acacia

When Renaissance era Latin scholars translated Avicenna's work during the 12th and 13th centuries in centers like Toledo and Salerno, they retained the word "Acacia" from "Aqaqia" which the Arabs previously extrapolated from the Greek "Akakia". How can a lobby group erase such an intra-civilizational heritage?

The Acacia by this point in the Renaissance was firmly associated with the gum bearing African thorn tree. This fixed the identity of "Acacia gum" in European medical language for centuries to follow [13]. By 1580, gum-acacia was firmly embedded in European pharmacy and had begun to appear in confectionery recipes. The Pharmacopoeia Augustana (Augsburg, 1580) lists "Succus Acaciae" among the mandatory simples. Parallel northern Italian antidotaries (e.g., Ricettario Fiorentino, 1597) repeat the entry, so by 1600 the drug belonged to the standard European dispensary [14].

The 16th century Renaissance saw the Mediterranean trade boom to new levels and an intensification of Acacia gum trade. Renaissance pharmacopoeias confirm its prominence: the Pharmacopoeia Londinensis (1618) names Acaciae Succus; the Pharmacopoeia Basileensis (1561) repeats Succus Acaciae; and the Antidotarium Florentinum (1498 ed.) prescribes Succus Acaciae in multiple electuaries, each text explicitly identifying the drug as the juice of the of the African thorn-tree [15]. Importantly, these early botanicals and material-medica literature began to supply a botanical portrait of the source tree described consistently as "spiny trees from Aegyptus, Aethiopia, and Barbariae," clearly ruling out the Australian continent, whose flora was unknown to Europeans before the late 17th century.

A 1653 British medical text, In Nicholas Culpeper's The Complete Herbal (London, 1653), the entry "Acacia (Gum Arabick)" states, "It is cold and dry in the second degree, exceeding binding; it stayeth all manner of fluxes of blood, healeth ulcers in the mouth, fasteneth loose teeth, and is good for spitting of blood." Adding that, "It is the gum of a thorny tree growing in Egypt and Arabia," thus confirming its African/Asian provenance [16].

Here we witness a linguistic lineage: Greek Akakia → Arabic Aqaqia → Latin acaciae succus → Renaissance pharmacopeias referring to "Acacia gum". The tree's reference remained constant: a thorny, resinous African tree, known to travelers, traders, and physicians for its powerful gum and its medicinal pods and leaves.

A Trade Documentation Boom

By the early 17th century, colonial mercantilism brought structured documentation to trade. The British East India Company and Dutch East India Company both listed Acacia gum as a traded item from Sudan, Sennaar, Abyssinia, and Egypt, with no mention of Australia's species. European botanists like John Ray cataloged the source of this gum and identified it with the African thorn trees [17].

These manuals therefore carried the African identity and astringent reputation of Acacia gum well into the 17th century, long before the first published descriptions of Australian Acacia species by Cavanilles (1797 – 1802) and Robert Brown (1813).
When Carl Linnaeus formalized binomial nomenclature in 1753, he retained the classical name Acacia for the gum-yielding thorn trees of

Egypt and the Nile (Acacia nilotica) thereby cementing the term that had circulated since the writings of Pliny and Dioscorides. Subsequent botanical codes have conserved A. nilotica as the type species of the genus [18]. Linnaeus himself listed Acacia nilotica as the type species for the genus, anchoring the classical tradition in a formal scientific framework.

This act by Linnaeus was not merely a convenience of classification; it was an acknowledgment of precedent. Linnaeus, a Latinist and bibliophile as much as a botanist, was aware of the classical texts. He did not pluck the name Acacia from thin air. He honored a lineage of botanical and cultural recognition that stretched from ancient Nubia and Egypt to Hellenistic Alexandria and early modern Europe. For nearly two and a half centuries after Linnaeus, the name Acacia referred first and foremost to the African trees that produced the historical gum of commerce which are Acacia seyal, and Acacia nilotica. Both were essential to the religious, pharmacological, and economic history of three continents.

When European exploration of Australia began in the late eighteenth century, Joseph Banks and Daniel Solander collected numerous "wattle" specimens but labelled them Mimosa in their field notes. The first botanists to publish Australian species as Acacia were Antonio Cavanilles (1797, 1802) and James Edward Smith (1797). Later, Robert Brown (1813), George Bentham (1864) and Ferdinand von Mueller (1887) retained the wattles in Acacia because their phyllodes, seedpods and flower structure matched the diagnostic characters of the African type species Acacia nilotica. Thus, their placement was morphological, not historical and definitely not practical usage. The Australian plants shared none of the long-documented medicinal or commercial uses that had defined "Acacia" (and its extracts) for two millennia [19].

As touched upon earlier, the Australian species were chemically distinct, medicinally inert, and often toxic. No historical use mirrored the African tree's uses. No sacred temple adorned its properties. No mummy was wrapped using its sap. No Avicennian healer prescribed it.

The Wattle Acacia Inclusion Dilemma

Fast forward two centuries later, this broader Wattle inclusion into Acacia led to a taxonomic crisis. In 1986, Queensland botanist Leslie Pedley unsheathed the taxonomic scalpel and performed a bold corrective surgery on a genus that had grown unwieldy. In a single paper, he restored the historic name Acacia to its African heartland, recognized the thorny, spicate-flowered relatives as Senegalia, and most daring of all, banished the entire Australian cohort to the resurrected genus Racosperma. Pedley's proposal wasn't mere tinkering; it was a principled stand for phylogenetic clarity and historical justice, delivered long before the nomenclatural firestorm that followed [19].

On the opposing side of Les Pedley was the Australian interest groups who wanted to steal the Acacia name and appropriate it for their native shrubs. Their justification was technical: most Australian species formed a monophyletic group based on DNA evidence, and, under the rules of botanical priority, a genus name could be "conserved" for the majority. Apparently, the fix was "purely scientific". Once chloroplast-DNA trees showed the phyllodinous Australian wattles huddled neatly in a single clade, the Acacia heisters planned and eventually made their move.

The Acacia heist started by invoking Article 14 of the ICBN code and ended with the airfreighting of the Acacia Name from Acacia nilotica (Egypt) to A. penninervis (Sydney). This heist story will demonstrate how through political lobbying, Acacia was "conserved" for the Australian wattle and stripped from the original Egyptian thorn tree [20].
But herein lay the problem. While Australians had more species by number (due to their taxonomic culture, discussed later in more detail), they had

less historical claim by every other metric (cultural, commercial, religious, and scientific). The African/Asian species had defined the name for millennia. They were the Akakia of Theophrastus, the succus acaciae of Dioscorides, the lignum acaciae of Jerome, and the Aqaqia of Avicenna. To strip them of the name in favor of botanically similar but culturally disconnected species was not scientific clarity; it was nomenclatural colonialism.

If nomenclature is to serve both science and society, it must be accountable to both. The case of Acacia is unique in its scope and symbolic gravity. It is not just a genus name; it is a cultural linguistic artifact, a living witness to our shared past. For more than 2,000 years, the name Akakia (later Acacia) referred to the African thorn trees that bled healing resin and built holy altars. They earned it in every sense: spiritually, medicinally, economically, and linguistically.

The Consequence of Taxonomic Anachronism

Today we are left with a dilemma. To preserve the truth of history and nomenclature; we must reject the modern classification that gives ownership of the name "Acacia" to species that have no traceable historical, medicinal, or cultural footprint until the 19th century. Australian wattles (now called Acacias) were unknown to the ancient world. They were unrecorded in:
- Egyptian temples
- Hebrew scriptures
- Bibles and other religious texts and their translations
- Greek medical writings
- Roman natural histories
- Islamic pharmacopoeias
- European merchant inventories before the 1800s

No Australian wattle appears in any textual corpus prior to the 18th century, a fact easily verifiable across major botanical and pharmacological databases. Australian explorers such as Joseph Banks and George Bentham only began cataloging these trees during the late 18th and 19th centuries [21]. While Banks admired their ecological value and wood properties, he made no claim of equivalence with the African Acacias known from global medicine.

Moreover, most Australian species such as A. pycnantha, aneura, and dealbata:

1. Do not exude edible or medicinal gum
2. Are rich in tannins and other toxic alkaloids
3. Have no history of global trade
4. Are largely used for ornamentation, nitrogen-fixation, and reforestation which historically was not attributed to the thorn tree
5. Most evident of all is that Australian wattles are given a name called the "Thorn tree" when in fact 98% don't have thorns at all!

In contrast, Acacia nilotica/seyal/tortilis are not only the true historical bearers of the name; but are also listed in modern pharmacopoeias, global trade agreements, and as food grade ingredients of prebiotic fiber significance. The consequence of this misnaming is historical and cultural erosion.

The reclassification creates a deep cultural and historical error because literature referencing "Acacia" and its applications for the past 2,300 years now appears to reference Australian trees, which is categorically false. A student or researcher might now assume Dioscorides or Avicenna referred to Australian plants which is a dangerous mistake in pharmacognosy. The study of civilization and the evolution of linguistics will now be tainted with name appropriation as millennia worth of trade, mythology, and medicine is now attached to the incorrect tree. The name "Acacia" belongs to a specific legacy, not to a botanical pen holder in Vienna.

An unknown wise man once said, "names aren't labels. They're stories". The name "Acacia" is a story at least 2,300 years in the making. A story that begins in the Nile, grows through the Mediterranean trade, flowers in Greek medicine, hardens in Islamic pharmacology, and becomes immortal in global commerce.

Akakia the thorn tree, is spiritually attached to: "not evil", "free from corruption". This was not a word chosen lightly. It was a testament to the tree's incorruptible wood, its healing gum, and its sacred role in rituals from Karnak to Jerusalem. To call a different tree by this name; a tree unknown to Theophrastus, unused by Avicenna, unrecognized by the authors of Exodus, is not merely inaccurate; it is a betrayal of the botanical record and a rewriting of cultural memory. To sever that name from the tree that built the story is not just misleading. It is unjust.

References:

1. Ḥunayn b. Isḥāq. Arabic De Materia Medica. Walters MS W.750, fol. 24r (c. 870 CE).
2. Ibn al-Bayṭār, Abū Muḥammad ʿAbd Allāh. Kitāb al-Jāmiʿ li-Mufradāt al-Adwiya wa al-Aghdhiya. Cairo: al-Maṭbaʿa al-Amīriyya, 1874, vol. 3, fol. 8.
3. al-Zahrāwī, Abū l-Qāsim. Kitāb al-Taṣrīf (Liber Servitoris). Edited and translated by Sami K. Hamarneh and Glen Sonnedecker. Leiden: Brill, 1963, 109–110; Ṣābur ibn Sahl. al-Aqrabādhīn al-Ṣaghīr. Edited by I. Dietrich. Leipzig: Teubner, 1887, § 114.
4. Avicenna (Ibn Sīnā). Al-Qānūn fī l-Ṭibb (The Canon of Medicine). Book II, Fen 2, chap. 21. Cairo: Būlāq Press, 1877, vol. 2, 65. Translated by Laleh Bakhtiar. Natural Pharmaceuticals, Great Books of the Islamic World, 2012, 55–57.
5. Rudge, A.J., and L.A. Street. "Acacia Gum: Chemical Characteristics and Uses." Journal of Natural Products 43, no. 2 (1980): 200–206.
6. Orchard, Anthony E., and Bruce Maslin. "Proposal to Conserve the Name Acacia." Taxon 52, no. 2 (2003): 362–363.
7. Goitein, S. D. A Mediterranean Society. Vol. 1: Economic Foundations. Berkeley: University of California Press, 1967, 362–365; Lucas, A., and J. R. Harris. Ancient Egyptian Materials and Industries. 4th ed. London: Edward Arnold, 1962, 301–2.
8. Kaur, Gurinder, et al. "Antidiarrhoeal Activity of Acacia nilotica Willd. ex Del. Pod Extract." Indian Journal of Pharmacology 37, no. 4 (2005): 167–171; Banso, Amina, and Samuel Adeyemo. "Evaluation of Antimicrobial Properties of Tannins Isolated from Dichrostachys cinerea and Acacia nilotica." African Journal of Biotechnology 6, no. 15 (2007): 1785–1787; Krishnamurthy, Praveen Kulkarni, et al. "Antioxidant Capacity of Acacia nilotica Leaf Extracts." International Journal of Pharmacy and Pharmaceutical Sciences 5, no. 3 (2013): 372–376.
9. Seigler, David S. "Chemical Constituents of Some Australian Acacia Species." Economic Botany 57, no. 2 (2003): 250–256; Everist, Selwyn L. Poisonous Plants of Australia. 2nd ed. Sydney: Angus & Robertson, 1981, 201–205.
10. Bereket, Hekîm. Tuhfe-i Mübârizî: Metin-Sözlük. Edited by Binnur Erdağı Doğuer. Ankara: Türk Dil Kurumu, 2013, 112–113; Çelebi, Emîr. Enmûzecü't-Tıbb. Istanbul University Library, MS TY 1925, fols. 62r–63r.

11. Mesue the Younger. Antidotarium sive Grabadin medicamentorum. Edited by D. R. Sudhoff. Leipzig, 1914, 72–73; The Dawn of Drug Safety. Uppsala Monitoring Centre, 2013, 9; Antidotarium Nicolai. Mellon MS 15, Yale University; Ibn Buṭlān. Tacuinum Sanitatis. Paris, BnF Lat. 9333, fol. 10r.
12. Anderson, David M.W. "Gum Acacia—Chemical and Industrial Aspects." Food Hydrocolloids 2, no. 3 (1988): 329–332.
13. van Dalen, Dorrit. Gum Arabic: The Golden Tears of the Acacia Tree. Leiden: Leiden University Press, 2019.
14. Pharmacopoeia seu Medicamentarium pro Republica Augustana. Augsburg, 1580, fol. 42v. Bayerische Staatsbibliothek, Res.2 Pharmaz. 241.
15. Ricettario Fiorentino di Nuovo Impresso. Florence: Lorenzo de' Morgiani, 1498, sig. A3r; Pharmacopoeia Basileensis. Basel: Apiarius, 1561, fol. 26v; Pharmacopoeia Londinensis. London: Thomas Snodham, 1618, 86.
16. Culpeper, Nicholas. The Complete Herbal. London: Peter Cole, 1653, sigs. A3v–A4r.
17. Ray, John. Historia Plantarum. Vol. 2. London: John Martyn, 1688, 1433–1434.
18. Linnaeus, Carl. Species Plantarum. Vol. 1. Stockholm, 1753, 521.
19. Pedley, Leslie. "Derivation and Dispersal of Acacia." Botanical Journal of the Linnean Society 92 (1986): 219–254.
20. Orchard, A., and B. Maslin. "Proposal 1584 to Conserve the Name Acacia." Taxon 52 (2003): 362–363; Miller, J., and R. Bayer. "Molecular Phylogenetics of the Acacia Alliance." Australian Systematic Botany 13 (2000): 21–46; Turland et al. "Nomenclature Section Report." Taxon 55 (2006): 795–814.
21. Mabberley, D. J. Joseph Banks: A Life in Botany. Berkeley: University of California Press, 2023; Flora Australiensis. Vol. 2. London: Lovell Reeve & Co., 1864.

CHAPTER 4

ACACIA UNITED – ACACIA DIVIDED

By the time European scientists began formalizing plant names in the 18th century, this tree, once revered in temple and tomb, was swept into a system struggling to keep up with global biodiversity. In 1753 Carl Linnaeus, the father of modern taxonomy, formally established the genus Acacia in his landmark Species Plantarum, classifying species such as the Egyptian thorn into Acacia nilotica [1]. A year later, Philip Miller embraced Linnaeus's classification in his Gardener's Dictionary (1754), furnishing practical horticultural advice under the heading "Acacia" and highlighting Acacia nilotica as the archetypal species [2].

This wasn't just a name. It was a resurrection. Miller tied the modern genus to the historical record. He etched the tree of Dioscorides and the Ark into the fabric of scientific classification. It was a taxonomic act of reverence.

Then came the storm.

The 18th and 19th centuries saw an explosion of global exploration. With it came a flood of plant specimens, Australian wattles, American shrubs, and African giants. Botanists scrambled to fit them into boxes. Enter George Bentham, a titan of botanical organization. Bentham saw chaos in the genus Acacia. He sought order, and in the process, gave it structure.

Between 1842 and 1875, Bentham reviewed hundreds of species and crafted a system grounded not in pod shape (which was variable and often missing in preserved specimens), but in floral anatomy, specifically the number and arrangement of stamens [3]. Where others saw confusion, he saw signal. His insight? True Acacia species had numerous free stamens. He saw diversity and even anticipated that Acacia might not hold up as a single genus for long.

Yet Bentham resisted carving up the genus. He had flirted with the idea of a separate genus (Vachellia) based on pod differences, but he recoiled at the last second. Splitting Acacia, he feared, would cause unnecessary fragmentation. He chose harmony over division, lumping where he might have split. This decision would echo for generations. For a century, Bentham's Acacia, sprawling but ordered, dominated botany. Unfortunately, harmony is rarely permanent.

As the 20th century progressed, a new wave of data emerged. Microscopy revealed differences in pollen grains. Cytologists mapped chromosome numbers. Chemists analyzed extract structures. The result is botanists began to suspect that Bentham's Acacia was not a unified whole but a polyphyletic chimera, a genus made up of unrelated branches masquerading as kin.

In the early 20th century only a handful of new acacia related genera were established. In 1928, Britton & Rose erected Acaciella for a group of Neotropical wattles formerly placed in Acacia; and in 1934, Auguste Chevalier separated Faidherbia for the unique African "apple-ring acacia" (Acacia albida) [4]. Most of these early segregates did not gain wide acceptance at the time and Bentham's broad Acacia still held sway. However, they presaged later splits: Faidherbia is now a recognized separate genus, and Acaciella would later be vindicated (under those very names) by molecular studies.

By the mid-20th century, the genus Acacia had grown into something of a botanical leviathan. Its branches extended across continents, its species numbering over a thousand, and its identity increasingly difficult to define. Acacia was no longer a singular genus, it was an empire, and like all empires, it teetered on the edge of collapse. Behind herbarium doors and inside dim lecture halls, a reckoning loomed.

Clues had been accumulating for decades. The Acacia genus, though held together by tradition and utility, was being pulled apart by evidence. Researchers peered through microscopes and saw pollen grains that bore no family resemblance. Chromosome counts, floral anatomy, and even seed structure pointed to divergence, not unity. The elegant taxonomic system that Bentham had built was quietly failing the test of modern scrutiny.

In 1972, Jacques Vassal conducts a detailed ontogenetic and pollen-morphology study of Acacia that clarifies characters shared and diverged among African, Asian, and Australian lineages, and although some claim he himself erected subgeneras Acacia, Aculeiferum, and Phyllodineae; This was not recorded in any primary taxonomic literature. Nonetheless his work provided credence for what was to come [5].

Leslie Pedley steps in to save the day

Then came the honorable Australian intervention. In 1986, Les Pedley, an Australian botanist, took Vassal's classification and made a bold move. He proposed formally splitting Acacia into three genera:

1. Acacia for the African/Asian thorn trees.
2. Senegalia for the other thorny species.
3. Racosperma for the Australian phyllode-bearing trees [6].

According to nomenclatural rules, the name Acacia should stay with the group containing Acacia nilotica, the original type species. That would mean African Acacias keep the name, and Australians adopt Racosperma.

Logical, historically grounded, and rule-abiding. Unfortunately, logic was no match for politics.

Pedley's split was not enacted at that time. The botanical community was reluctant to adopt such a sweeping change, largely because it would cause massive nomenclatural upheaval as hundreds of Australian "wattles" would need new names (as Racosperma), and many African species would shift to Senegalia. The Australian interest groups were the biggest opposers to Les Pedley and they quietly planned in the background.

Acacia divided

The most damning verdict came from DNA tests. By the 1990s, molecular phylogenetics (a new scientific lens) began to uncover what the eye could not. Acacia, as then defined, was polyphyletic. Its species did not descend from a common ancestor. Instead, Acacia housed five separate evolutionary lineages. What looked like unity was an illusion born of convergent evolution and old habit.

By the turn of the 21st century, comprehensive molecular phylogenies started making breakthroughs. In 2002 Miller, Murphy & Bayer demonstrated that Acacia sensu lato is polyphyletic, comprising at least five distinct clades [7]. These lineages broke down as follows:

1. The Australian phyllode-bearing trees, formerly Bentham's Phyllodineae, numbering nearly 1,000 species.
2. The African and Asian thorn trees, including A. nilotica, A. seyal, and A. tortilis, all part of Bentham's Gummiferae.
3. The other thorny acacias like A. senegal, A. mellifera, and A. pennata, echoing Bentham's Vulgares and Filicinae.
4. A group of delicate, thornless acacias from the Americas.
5. A smaller lineage of North American species with thickened roots and long flower spikes.

There was also Faidherbia, the genus to which Acacia albida had long been reassigned, and a few stragglers that defied neat categorization. The central truth was clear: Acacia, as a single genus had become scientifically untenable. The question now was political: Who would keep the name?

By the rules of the International Code of Botanical Nomenclature, the answer should have been straightforward. The name Acacia would remain with the group containing the original type species, Acacia nilotica. This meant the African Acacias; iconic, storied, and historically documented, would keep the name. The Australians, who possessed the most species but none of the name's historical weight, would rebrand their trees under the older, lesser-known genus name Racosperma. Unfortunately, what was to come was more than taxonomy.

Acacia had become more than a genus. In Australia, it was a national emblem. The Golden Wattle (Acacia pycnantha) adorned crests, uniforms, and postage stamps. The tree featured in poetry, parades, and identity. Renaming it to Racosperma pycnanthum was not merely inconvenient, it was unthinkable. So, the Australian interest groups made their move.

References:

1. Carl Linnaeus, Species Plantarum, vol. 1 (Stockholm: Laurentius Salvius, 1753), 382; vol. 2, 521.
2. Philip Miller, The Gardener's Dictionary, 7th ed. (London: Printed for the Author, 1754).
3. George Bentham, "Notes on Mimoseae, with a Description of a New Species from Nova Guinea," London Journal of Botany 1 (1842): 488–512; Flora Australiensis: A Description of the Plants of the Australian Territory, vol. 2 (London: L. Reeve & Co., 1864), 347–532; "A Revision of the Mimosaceae, with Special Reference to the Flora of Australia," Journal of the Linnean Society, Botany 13 (1875): 1–568.
4. N. L. Britton and J. N. Rose, "Acaciella," North American Flora 23 (1928): 96; A. Chevalier, "Faidherbia, gen. nov.," Revue de Botanique Appliquée et d'Agriculture Coloniale 14 (1934): 876.
5. Jacques Vassal, "Apport des recherches ontogéniques et séménologiques à l'étude morphologique, taxonomique et phylogénique du genre Acacia," Bulletin de la Société d'Histoire Naturelle de Toulouse 108 (1972): 105–247.
6. Leslie Pedley, "Notes on Acacia and Related Genera (Leguminosae: Mimosoideae)," Austrobaileya 2, no. 3 (1986): 315–317.
7. J. T. Miller, D. J. Murphy, and R. J. Bayer, "Phylogenetic Relationships of Acacia s.l. (Fabaceae: Mimosoideae): Evidence from Chloroplast DNA matK and Nuclear Ribosomal ITS Sequences," Australian Systematic Botany 15, no. 1 (2002): 17–29; B. R. Maslin, M. D. Crisp, and J. J. Ladiges, "Molecular Phylogeny of Acacia s.l. (Leguminosae: Mimosoideae) in Australia: Evidence for Five Major Clades," Australian Systematic Botany 16, no. 1 (2003): 1–18.

CHAPTER 5

LES PEDLEY SPARKS THE "ACACIA WARS"

In the world of scientific taxonomy, where debates typically unfold in dusty journals and behind herbarium doors, the "Acacia Wars" were something else entirely. It was a full-blown geopolitical chess match masquerading as botanical housekeeping. Beneath the Latin binomials and procedural motions lay a campaign fueled by national identity, economic ambition, and a deep understanding of how to win without appearing to fight. At the center of this drama was Australia, an unlikely underdog that transformed a defensive position into an aggressive, well-planned conquest of a name that was culturally and historically associated with Africa for over two millennia.

To understand just how strategic this operation was, one must go back not to 2003 (when the infamous Orchard & Maslin proposal 1584 was submitted), but to 1986 when Les Pedley first proposed a split of the oversized and polyphyletic Acacia genus [1]. His plan? Retain the name Acacia for the African lineage, the very trees known throughout antiquity for producing the famed Acacia gum. In essence, it was an Australian taxonomist who fired the first shot, against Australian nationalistic interests.

This original proposal was not an oversight; it reflected the historical reality that the African/Asian species had the two-thousand-year-old established and well-documented cultural, medicinal, and economic legacy. The name Acacia evolved from Egyptian temples, Biblical altars, Greco-Roman medicine cabinets, Islamic pharmacopoeias, and recent trade documentation. By contrast, the Australian wattles lacked any of that heritage.

The Problem is that Pedley's proposal was a taxonomic revision, not a nomenclatural act with binding force. For such a major reclassification to be adopted internationally, it needed endorsement or implementation by the key global authorities: the Nomenclature Committee for Vascular Plants (NCVP) and ultimately the International Botanical Congress (IBC). Pedley's paper was a scientific argument, not an official proposal to the nomenclature committees. It remained a published suggestion until acted upon by wider consensus

Some argue that Les Pedley's proposal in 1986 was never implemented due to "regulatory inertia" and the fact that it was never formally brought before the Nomenclature Committee for Vascular Plants (NCVP) or the International Botanical Congress (IBC), a significant oversight by African and Asian botanists [2]. The fact that two IBC events were executed, in Berlin (1987) and Tokyo (1993), before the controversial Vienna IBC in 2005 attests to the sheer complacency of the Africans and Asians interested in preserving the name.

Pedley's proposal was primarily taxonomic and may have lacked the time, funding, and administrative support to navigate the lengthy Code process. This was especially true because it was done in an era before online submissions and consolidated herbaria networks. The years will also show that Australian botanists led a "quite resistance" to Les Pedley's split and started working on their own conservation strategy. African complacency gave them two decades to perfect their heist.

What's certain is that Australian institutions did not publicly contest the science of the split. Instead, they publicly embraced it and then rewrote the rules of engagement in the back offices of international institutions. Rather than opposing the division, they redirected its consequences. By the late 1990s, botanists like Bruce Maslin (Western Australian Herbarium) and Tony Orchard (Australian National Herbarium) were not only researching Acacia phylogeny, but also actively developing proposals to retain the name Acacia for the Australian wattle contingent. Their approach was legalistic, methodical, and astonishingly foresighted.

The "Acacia Wars" were remarkable in that they drew in not just scientists but also national pride and economic considerations. Key individuals and institutions took unmistakable sides. On the Australian interest group side, the campaign was led by figures like Bruce Maslin (Western Australian Herbarium) and Tony Orchard (Australian National Herbarium), co-authors of the 2003 proposal 1584 (to grant Acacia to the Australian Wattle trees) [3]. They were supported by many Australian systematists and backed implicitly by Australian institutions that valued Acacia as part of the national heritage. Australia's government botanists and herbaria had a stake as Australia's national floral emblem is the Golden Wattle (Acacia pycnantha), featured on the coat of arms and widely celebrated.

The 2003 proposal to reassign Acacia away from its original African type to an Australian species was submitted to the International Association for Plant Taxonomy (IAPT) as proposal 1584. This was no academic footnote; it was a preemptive strike. The language was surgical, the rationale procedural, and the intention clear: keep Acacia Australian, even if the African species had the historical claim.

While Australian institutions led the charge to conserve *Acacia* for their native wattles, the counter-voice came primarily from global legume specialists, particularly from Africa and South America. Tensions grew when procedural irregularities at the 2005 Vienna IBC allowed the proposal to pass despite majority opposition, a situation later criticized as "minority rule" [4].

What most outsiders didn't see was the infrastructure that had been quietly constructed behind this campaign. In opposition to Les Pedley's anti-nationalistic Acacia solutions; Maslin and his lobby launched the WorldWideWattle project between 2002-2003. The website that would serve as the digital stronghold for Australia's position. Sponsored by CSIRO and other national scientific bodies; the organization provided a centralized repository of research, nomenclatural updates, and strategic messaging. It didn't just present data, it controlled the narrative. A section on the "Acacia name issue" kept the global community updated with Australia's framing of the debate, softening the ground for the coming vote.

It is not hard to see the patriotic motive as losing the genus name for your national flower would be symbolically jolting. As Orchard and Maslin highlighted, Acacia pycnantha itself would change to an alien name (Racosperma pycnanthum), requiring reprinting of countless documents and signage. Australian delegates stressed that their country's identity was entwined with wattles; Australia even celebrates "Wattle Day" each year. It is evident that this sentiment likely resonated in their lobbying.

They say it was just a botanical tussle, but beneath the veneer of Latin names and herbarium labels, what was to come was a full-blown campaign of national intrigue. In one corner stood Australia's golden legion (Bruce Maslin and Alan Orchard) quietly drafting their 2003 masterstroke to hijack the name Acacia for the continent's own. Behind closed doors, state herbaria chiefs whispered about safeguarding a "national heritage" worth billions in forestry and agriculture. This wasn't a mere clerical adjustment; It was a strategic move with global consequences, effectively rewriting the taxonomic map of one of the world's largest and most ecologically important genus.

Australia has nearly 1,100 Acacia species covering vast landscapes. The

Australian Wattles are significant for land restoration, timber, tannin, cut flowers, and other, similar uses. Several Australian wattles (e.g. Acacia mangium, A. mearnsii) are grown in plantations across Asia and Africa for wood and pulp. Maintaining a consistent name was argued as beneficial for forestry and trade. The Australian delegation argued that renaming these to Racosperma would cause confusion in forestry programs worldwide.

Some critics later speculated that Australia was also keen to secure the "Acacia" brand for any current or future commercial products derived from its wattles (from wattle seed food products to potential gum or extracts) without competition from African Acacia retaining the name [2]. While not explicitly stated at the Congress, protecting national branding was certainly an undercurrent.

The preemptive attack was launched as proposal 1584 in May 2003. In the coming months the Committee for Spermatophyta began its deliberations on Proposal 1584. By December the interest group's efforts had already begun to convince undecided delegates on the International Botanical Congress' Nomenclature Committee.

In December 2003, as the Committee for Spermatophyta began its deliberations on Proposal 1584 (submitted by Maslin 7 months earlier), Les Pedley, the silent hero, launched a counterattack. With his unwavering allegiance to the rules of botanical nomenclature, he dropped a bombshell: an 834-entry reclassification of Australian Acacia species into the genus Racosperma [5]. Published in Austrobaileya, his monumental paper "A Synopsis of Racosperma C. Mart. (Leguminosae: Mimosoideae)" was a taxonomic act of resistance. Les Pedley's publication was not just a scientific statement, but a challenge to what he saw as a growing tide of political maneuvering that threatened to override the very principles taxonomy was built on.

Les Pedley's publication was a well-timed, precision strike. Pedley saw the Maslin-Orchard proposal for what it was: a geopolitical land grab disguised as nomenclatural housekeeping. In response, he did the only thing a principled taxonomist could. He resurrected Racosperma, a name first proposed by Carl Friedrich Philipp von Martius in 1829, and filled it with names (800+ of them) to give the African type species its rightful taxonomic dominion [5]. He launched his counterattack strategically while the Committee for Spermatophyta was deliberating so that it would be the most relevant/recent deciding factor. R.A. Brummitt, reporting for the Committee in 2004, called the timing "surprising" and acknowledged the combinations were now incontestably available [6]. Even still, lobbying momentum already favored the conservation proposal 1584.

Why did Pedley wait seventeen years to publish the mass combinations that the interest group claimed were missing (giving credence to the Australian numerical argument)? The answer might lie in cost–benefit logic, not hesitation. At the time, large-scale combination papers demanded painstaking Latin diagnoses, manual type citations, and a sliver of readership. Publishing 800 names when the community had not asked for them would have been intellectual noise. Only when the "too few combinations" argument became decisive did the publication become strategically valuable. Pedley's December 2003 blitz was not a change of heart but a measured intervention; removing the final factual barrier of keeping Acacia with the original thorn tree and letting principles decide.

Had this been a normal scientific process, Pedley's action would have changed the conversation. A reclassification of this scale and precision would have demanded debate, if not full reconsideration. Suspiciously, it was met with something else entirely: bureaucratic silence. The committee acknowledged the publication and then dismissed it. One member famously remarked, "We might as well accept it." In support of Les Pedley's solution but the vote reflected that he was outnumbered [6].

Inside the Committee for Spermatophyta, politics was already in motion. Despite Pedley's massive reclassification, the proposal to conserve Acacia for Australian species gained traction. In their official report published in Taxon in August 2004, the committee referenced some of the discussions that took place as they were deliberating whether to approve Maslin's proposal or not: "Surprisingly, around 800 new combinations in Racosperma for the Australian species were made in one paper in December 2003 in the middle of the committee's deliberations...", "now that this has been done, we might as well just accept the situation and recommend taking up Racosperma", "Others do not see the formal publication of these names as a critical factor...". These quotes testify that some committee members were already hell bent on granting Australia the Acacia name and what happened next proves this [6].

Sensing that his initial intervention did not penetrate the thick lobbyist membrane guarding the committee, Pedley followed up with another paper in March of 2004 titled, "Another view of Racosperma" in the Acacia Study Group Newsletter No. 90 [7]. It was a heartfelt, rule-based plea against what he saw as a taxonomic coup. Pedley argued that retypifying Acacia would not only break with nomenclatural tradition but sow chaos across scientific databases, conservation records, and herbaria worldwide. Yet again, the response from the other side was muted.

According to the Vienna Code's official procedures (Regnum Vegetabile 146, Appendix III), any validly published combinations and typification proposals must be considered on formal grounds by the Committee [8]. Pedley's reclassification of over 800 Acacia species under Racosperma in 2003, and his follow-up in March 2004, both met these criteria. Yet despite their procedural validity, the Committee took no formal action to adopt or even engage with them. The disconnect between the codified process and what actually took place shows clearly that lobbying, not taxonomy, played the decisive role.

In the meantime, you would expect Maslin and his allies to quickly respond to Les Pedley's two publications as the Committee is deliberating a pivotal decision regarding Proposal 1584. However, it seems Maslin was so confident about the committee approving his proposal that he did not even bother responding to neither Les Pedley's 800 combination publication in December 2003 nor to his follow-up publication in March 2004. Instead, he waited until November 2004 to respond to Les Pedley, five months after the committee approved his proposal [9]. In June 2004, the Committee voted: nine in favor, six against [6]. Just like that, decades of nomenclatural tradition shifted. The vote was officially reported two months later, in Taxon Volume 53, Issue 3.

Having secured the Committee approval of Proposal 1584; Maslin and Orchard finally responded to Les Pedley in the Acacia Study Group Newsletter No. 93. However, this wasn't an invitation to debate, it was a rebuttal. The Maslin response to Les Pedley was a bureaucrat's counterpunch to a purist's cry for order. It spoke volumes. Pedley, fighting with rules and publications, had brought a textbook to a public relations fight. His opponents brought votes, strategy, and the machinery of institutional power. Pedley's taxonomic bomb had not slowed the convoy or changed its direction. The destination had already been reached.

This initial Committee decision was only the first step in the coup d'état launched in Vienna in July of 2005 where delegates would vote to keep this committee decision or revoke it. In hindsight, the African and Asian response was fragmented, underfunded, and reactive. Botanists from South Africa, Ethiopia, and India voiced their outrage in papers and petitions, but without the institutional or political machinery to turn protest into policy. They lacked the funding, unified lobbying apparatus, and centralized media platform the Australians employed. While African botanists invoked cultural legacy and nomenclatural principles, the Australians delivered white papers, delegates, and procedural precision.

Politically, the Australians mobilized local allies and ensured high visibility at key venues. African botanists, in contrast, lacked the means to fund mass attendance or to host a counter congress. Their appeals to fairness, while valid, did not survive the procedural machinery. The Australians interest group understood this asymmetry and fully exploited it.

This preparation climaxed at the 2005 International Botanical Congress (IBC) in Vienna, where the Nomenclature Section gathered to vote on the proposal. Here, Australia's logistical strategy became apparent. Reports indicate that the Australian contingent flew in dozens of sympathetic delegates from Oceania, Southeast Asia, and elsewhere, an effort critics called "stacking the vote". The optics were dramatic: 55% of the delegates voted against the proposal, but due to a freshly new procedural requirement for a 60% supermajority to overturn a committee recommendation, Australia's minority vote prevailed, and the heist succeeded. This controversial new 60% supermajority rule was literally enacted by the section body right before the Acacia vote and while the 2005 IBC was in session.

This was not a victory of science; it was a victory of strategy. Australia had used the rules as weapons and won. In the next chapter, we discuss the details of what happened during the 2005 Australian Lobby's Vienna coup.

References

1. Pedley, Leslie. "Notes on Acacia and Related Genera (Leguminosae: Mimosoideae)." Austrobaileya 2, no. 3 (1986): 315–316.
2. Luckow, Melissa, Colin Hughes, Barabara Schrire, et al. "Acacia: Consequences of Typification Choices." Taxon 54 (2005): 513–519.
3. Orchard, Anthony E., and Bruce R. Maslin. "Proposal to Conserve the Name Acacia (Leguminosae: Mimosoideae) with a Conserved Type." Taxon 52, no. 2 (2003): 362–363.
4. Moore, Gerry, Daniel J. Murphy, and Peter H. Weston. "The Acacia Controversy Resulting from Minority Rule at the Vienna Nomenclature Section." Taxon 60 (2011): 852–857.
5. Pedley, Leslie. "A Synopsis of Racosperma C. Mart. (Leguminosae: Mimosoideae)." Austrobaileya 6, no. 3 (2003): 445–496.
6. Brummitt, R. K. "Report of the Committee for Spermatophyta: 55." Taxon 53, no. 3 (2004): 827–829.
7. Pedley, Leslie. "Another View of Racosperma." Acacia Study Group Newsletter, no. 90 (March 2004): 10–12.
8. International Code of Botanical Nomenclature (Vienna Code), 2006. Regnum Vegetabile 146. Appendix III, "Procedures and Guidelines for Nomenclature Sections."
9. Moore, Gerry. "Acacia: The Case Against Conservation." Bothalia 38, no. 2 (2008): 201–209.

CHAPTER 6

THE VIENNA COUP

As explained in the previous chapter, The first score of the Acacia heisters was in August 2004 when they convinced the Nomenclature Committee for Vascular Plants (NCVP) to disregard Les Pedley's taxonomic publications and vote in favor of recommending the appropriation of Acacia for the Australian wattles instead of the original African thorn tree. The result of which was formally reported in Taxon, Volume 53, Issue 3, in August 2004 [1]. This first score could have been argued as a scientific failure of conscience. However, the heisters did not stop there in their string of scores. Their second score was even more critical.

By the time the 17th International Botanical Congress (IBC) convened in Vienna in July 2005, the battle over the name Acacia had already entered a controversial phase. The controversy lay in the approval of a new rule that required a 60% supermajority (instead of 50%) to overturn the recommendation made by the "General committee". This meant that the previous 2004 recommendation (Acacia for the Australian wattles instead of the African thorn tree) by the committee would now need a 60% supermajority to overrule it instead of the 50% that was previously sufficient to overturn committee recommendations.

If this is not controversial enough, this rule was enacted during the IBC congress itself to ensure there was minimal time for anyone to do anything about it. They invented a new 60% rule, during the same Congress where Acacia was up for decision and put it into effect right away [2]. This cannot be interpreted as anything other than a traditional "Fait acompli" consistent with any coup. This procedural change, enacted just before the

Acacia vote, and while the IBC was in session fundamentally altered the balance of the decision-making power within the IBC process. The Australian heisters pulled their second score in the form of a coup d'etat.

While the new rule change was technically announced through distributed printouts. It was not widely publicized in advance and many attendees (including some seasoned taxonomists) were unaware of its practical consequences until its invocation during the Acacia vote. Many of the delegates who were aware of the new 60% supermajority rule, were not sure whether the 60% threshold applied to the "for" or the "against".

With the de facto rule in place and confusion rampant. The 450 delegates were asked to vote for or against the earlier 2004 recommendations. The tally results were very promising. 247 delegates votes "Yes" to overturn the recommendation (in doing so keeping Acacia with the African thorn tree) while 203 delegates voted "No" and thereby supported the recommendation to appropriate Acacia for the Australian wattles [3]. The majority have spoken, but it didn't matter this time. The 55% majority fell short of the new 60% supermajority rule.

A ripple of confusion spread throughout the hall. Delegates scrambled for their printed rules, flipping back and forth between the freshly voted-on procedures and the thick, impenetrable language of the Code. Was the chair really rewriting the rules on the fly? As the session chair explained the new interpretation, there was considerable confusion and surprise among the delegates. Many referred back to the procedural rules just adopted, seeking clarification, while others expressed concern that the process was being altered in real time [4]. The frustration among delegates was mainly that committee members had almost a whole year (since NCVP recommendation of August 2004) to transparently communicate any new rules. They chose not to do that and instead surprised everyone during the vote.

To be clear, this abrupt rule change was not announced before the IBC. There was no floor debate, and no verbal warning. Most delegates, still clutching their annotated codes, were blindsided. Some senior officials and secretariat members may have had forewarning, but the great majority of the delegates were left in the dark. This is not how such a pivotal decision was supposed to be handled. Per the code, all delegates should have had the opportunity to question, object, or seek clarification before any change in voting rules was enacted.

The rule change is described as a serious breach of expected procedural transparency. The scientific assembly, governed by the transparent and precise language of the International Code of Botanical Nomenclature, unfolded into a drama worthy of a detective novel, with plot twists that would echo through the annals of botanical governance. This was not how global nomenclature was supposed to be decided. Gerry Moore began piecing together the clues as soon as he returned home. In Bothalia and Taxon, Moore and his colleagues painstakingly reconstructed the vote [5], exposing the sleight of hand that had allowed Proposal 1584 to pass.

The inconsistencies deepened. In the published Vienna Code, the Acacia entry appeared without the usual asterisk reserved for names accepted by less than 60%, a subtle erasure, perhaps, but one that masked the historical anomaly from future botanists leafing through the Code's pages [5].

The repercussions rippled outward. Delegates left Vienna in disbelief. Many only realized the full implications after reviewing the records in the days that followed. The episode sparked widespread calls for reform. Critics argued that this breach undermined trust in the process, setting a precedent where the will of the minority could override the consensus of the global botanical community.

The 2005 Acacia vote was marred by accusations of political maneuvering and lobbying tactics that stretched far beyond the norms of academic discourse. Australian proponents, particularly Tony Orchard and Bruce Maslin (the authors of the conservation proposal 1584) reportedly made deliberate efforts to sway the outcome. According to multiple sources, they directed efforts to post dozens of supportive letters from Australian citizens on bulletin boards at the Vienna Congress, signaling widespread public backing for conserving Acacia for the Australian lineage [6].

At the same time, Australia's well-funded herbaria and botanical institutions ensured a disproportionately large number of registered voting delegates at the Congress. By contrast, delegations from Africa and Asia were less numerous and less organized due to financial and logistical constraints.

This disparity led some observers to accuse the Australian side of "stacking the vote" in their favor through a combination of national funding, preparation, and procedural exploitation [4]. The result, though procedurally valid, left a lasting impression that the outcome was shaped as much by campaigning and lobbying as by nomenclatural merit. The controversy remains one of the most politicized chapters in botanical naming history.

Defenders of the Vienna process, including respected figures like John McNeill and Nicholas Turland, countered that the decision was "technically legal". Division III of the Code allowed the Section to set its own voting procedures, and, they argued, delegates had (perhaps unwittingly) accepted the rule that committee-endorsed proposals needed 60% opposition to fail. Yet even these defenders admitted that this outcome was "probably not one to be adopted again", a tacit acknowledgment of its dubious legitimacy [7].

The reverberations echoed through the years. When the next IBC convened in Melbourne in 2011, opponents tried to force a re-vote, proposing that Acacia should be removed from Appendix III unless it achieved a true, majority vote. Unfortunately, the motion was ruled out of order. Quietly, the Melbourne Code restored the asterisk beside Acacia, an implicit admission that the 2005 outcome had broken with tradition and perhaps with justice [8].

Reading about this heist unfold; It's clear the Australian Interest group did not actually succeed through disputing botanical rules; they succeeded by changing the rules of engagement to secure their desired income. The baffling aspect is not the heist attempt; it's how a prestigious international organization gave away the keys and the safe for heisters to steal a 2,300-year-old name. That, perhaps, is the central lesson. In the global politics of taxonomy, the winner is not always the one with the oldest claim, but the one who plays the heist game.

The minority-rule incident galvanized reforms, leading to stricter voting procedures in subsequent Congresses. Now, any proposal to conserve or reject a name must achieve a true 60% affirmative vote, regardless of committee endorsements, a direct legacy of the Acacia controversy [4]. Had this been implemented in the Acacia vote, it would mean that the 60% threshold burden falls on the side that is recommending the proposal not overturning it.

In the aftermath of Vienna, SANBI publicly lamented the loss of a name it considered irrevocably African, accusing Australia of effectively "stealing" an emblematic term from the continent [9]. Although delegates cited "nomenclatural stability" to justify their vote, the sentiment among African voices was clear, this felt like the culmination of a painful severance, I call it a professional heist.

The Vienna coup stands as a cautionary tale: a moment when rules bent to pressure, transparency gave way to expediency, and global scientific heritage was outmaneuvered not in the pursuit of truth, but in the calculus of numbers and influence. It was a conspiracy played in daylight, written not in whispers but in the footnotes of a code. Its implications are still unraveling for a world of botanists, traders, regulatory bodies and historical experts.

The Australian campaign didn't stop there. After the Vienna vote, Australian herbaria quickly adopted the decision, updating the national names database as quickly as possible. Their rapid compliance suggested something more than responsiveness; it suggested preparation. Meanwhile, Maslin and Orchard continued to flood the literature with supportive arguments, including rebuttals to critics in Taxon. They weren't just defending a name; they were constructing a scholarly firewall to protect the outcome from future challenges.

In the years between the 2005 Vienna Congress and the 2011 Melbourne Congress, Australian taxonomists played a subtle but decisive hand in the long-running struggle over the name Acacia. Their most effective gambit appeared almost innocuously in April 2006, when botanists Phillip Kodela and Peter Wilson published a short, technical article in the journal Telopea. In just over ten pages they transferred 32 thorn tree wattles from Acacia into the resurrected genus Vachellia. They quickly implanted names such as Vachellia clarksoniana, V. ditricha and V. suberosa that were now, by the rules of botanical nomenclature, immediately legitimate and citable [10].

That slim paper accomplished far more than a routine tidying of Latin diagnoses. Under the International Code of Nomenclature, a name validly published in a peer-reviewed venue gains instant legal standing; no committee can veto it retroactively. By creating what lawyers call a fait accompli, Kodela and Wilson raised the administrative price of reversing

Vienna. Every herbarium sheet, seed bank permit, and state weed schedule updated after 2006 now bore Vachellia. A second U-turn in Melbourne would therefore mean pulping five years of labels and rewriting countless databases, a prospect most botanists' dread.

The timing also shaped perceptions of momentum. Australian advocates could claim that they had already finished the heavy lifting as their half of the global flora was neatly relabeled, while African opponents of the Vienna decision had yet to publish a matching set of Vachellia combinations for their own species. Briefing notes circulated among Congress participants in 2009 emphasized that "all relevant Australian names now exist in Vachellia," warning that a retrenchment would "generate yet another avalanche of paperwork" [4]. To undecided voters, many of whom cared far more about practical stability than about nationalistic symbolism, Australia's position began to look like the safer, less disruptive choice.

In the end, what happened in Vienna was not unforeseen, it was foretold. As early as 2005, Luckow and colleagues warned that typification was never just a taxonomic housekeeping matter; it was a decision with sweeping consequences for global nomenclature, ecological communication, and botanical justice. They cautioned that conserving Acacia for the Australian lineage, against its historical type, risked undermining the integrity of the Code itself. Their conclusion revealed that the potential consequences of such a choice extend far beyond the genus itself and threatens the credibility of nomenclatural governance [11]. In hindsight, their warning reads less like academic musing and more like a prophetic indictment of what would come to pass.

It wasn't just Acacia that was retyped, it was the entire principle of international scientific consensus that was redrafted under pressure. By the time delegates gathered in Melbourne in July 2011, the psychological weight of those 32 names had grown. Five years of routine usage had normalized Vachellia in floras, field guides and quarantine regulations. Proponents continuously argued that the pragmatic view was to keep things as they are [12]. Many observers later concluded that the quiet publication in Telopea had done more to coronate the Australian victory than any speech or lobbying campaign. What looked like a minor housekeeping exercise was in reality a masterclass in procedural strategy: act early, eloquently argue your questionable fact, entrench within decision making process, plant your initial success very deep, and then let the cost of reversal work for you.

In botanical chess, sometimes the smallest pawn moves (thirty-two names tucked into an unassuming Telopea journal) tips the entire endgame. The coming Melbourne outcome was sealed before it started. Even without the home advantage, the Australian lobby had already moved their checkmate.

References

1. Brummitt, R.K. "Report of the Committee for Spermatophyta: 55." Taxon, vol. 53, no. 3, 2004, pp. 827–829.
2. Moore, G. (2008). Acacia: The case against conservation. Bothalia, 38(2), 201–209.
3. McNeill, J. et al. (2006). "International Code of Botanical Nomenclature (Vienna Code)," Regnum Vegetabile 145
4. Moore, G. et al. (2011). "The Acacia controversy resulting from minority rule at the Vienna Nomenclature Section." Taxon 60: 852–857.
5. Moore, G. "The handling of the proposal to conserve the name Acacia with a conserved type." Bothalia, vol. 40, no. 2, 2010, pp. 185–190.
6. Moore, G., Murphy, D. J., & Weston, P. H. (2011). "The Acacia Controversy Resulting from Minority Rule at the Vienna Nomenclature Section." Taxon 60(3): 852–857.
7. McNeill, John. "XVII International Botanical Congress: Summary Report of the Nomenclature Section (Vienna 2005)." Botanical Electronic News, no. 356 (2005)
8. Wiersema, John H., Nicholas J. Turland, John McNeill, and Sandra Knapp. "Report on Botanical Nomenclature—Melbourne 2011: A Review of the Nomenclature Section at the XVIII International Botanical Congress." Phytokeys 41 (2014): 1–78
9. SANBI. Acacia for Africa! (December 2005), pp. 178, Veld & Flora
10. Kodela, Phillip G., and Peter G. Wilson. "New Combinations in the Genus Vachellia (Fabaceae: Mimosoideae) from Australia." Telopea, vol. 11, no. 2, 27 Apr. 2006, pp. 233-244.
11. Luckow, M. et al. (2005). "Acacia: consequences of typification choices." Taxon 54: 513–519.
12. Thiele, Kevin, et al. "The Controversy over the Retypification of Acacia Mill. with an Australian Type: A Pragmatic View." Taxon, vol. 60, no. 1, Feb. 2011, pp. 194-198.

CHAPTER 7

THE MELBOURNE CORONATION

If Vienna was the coup, Melbourne was the coronation. For those watching the Acacia saga unfold from the sidelines, the 2011 International Botanical Congress (IBC) in Melbourne looked like just another procedural stop in the world of taxonomic housekeeping. However, for those who understood the stakes, it was clear Melbourne was where the final seal would be applied to one of the most audacious nomenclatural heists in botanical history.

To fully grasp what unfolded in Melbourne, one must understand that the IBC does not simply land in a country by luck or rotation. Hosting rights for such an event require multi-year planning, academic coordination, government support, and substantial logistical capacity. Australia had all four. The fact that the next major vote on Acacia would be held on Australian soil, just six years after the controversial Vienna decision, was either a stroke of masterful planning or a geopolitical coincidence too large to ignore.

But coincidence seems unlikely.

The process for selecting IBC host cities involves formal bidding and lobbying within the International Association for Plant Taxonomy (IAPT). The final decision is made via a vote at the General Assembly at the end of each Congress, with voting rights extended to accredited delegates and officials from previous IBCs [1]. It is a system ripe for those with organizational muscles to mobilize resources, craft a persuasive pitch, and secure alliances. Australia, even before securing the Vienna decision in its favor, submitted a letter of interest to host the 18th IBC early 2005. They were notified of their successful bid during the 2005 IBC-Vienna [2]. The campaign to prepare the next IBC Congress home began after Australia confirmed that "proposal 1584" was both approved by the NVCP and the General committee. Australia securing the bid to host the next IBC in Melbourne Australia counts as their third successful score in this Grand heist.

Coincidences? I think not.

From the moment Proposal 1584 was tabled in 2003, the contours of a meticulous campaign began to take shape, less a matter of scientific consensus than of calculated maneuver. Committee pathways were expertly navigated, while Lis Pedley's taxonomically grounded publications (formally valid, published and scientifically binding) were conspicuously sidelined to clear the way for endorsement of the proposal. No sooner had the committee's approval been secured and we see Australia preparing to host the next International Botanical Congress, staking out home turf for the decisive act.

Then came Vienna, 2005. In a move as opaque as it was pivotal, a novel 60% supermajority rule was introduced mid-session. This unprecedented backdoor move rewrote long-standing voting precedent in real time. Though a clear majority of delegates (247 to 203) voted against conserving Acacia for the Australian lineage, the new procedural threshold flipped the outcome. Interestingly, in the same breath, Australia's bid to host the 2011 Congress was confirmed.

By the time Melbourne convened, the groundwork was complete. Australian taxonomists had already transferred many native wattles into Vachellia and Senegalia, with warnings that any reversal would unleash a bureaucratic avalanche. Every maneuver (taxonomic, procedural, and diplomatic) appeared choreographed to perfection. To observers across the botanical world, what unfolded no longer resembled the neutral arbitration of scientific nomenclature. It bore the unmistakable precision of a nomenclatural heist.

Once Melbourne was locked in, the battlefield shifted from procedural rulebooks to psychological advantage. This wasn't just about a vote anymore; it was about securing a narrative, managing visibility, and defusing opposition before it arrived. From 2005 to 2011, Australian scientists doubled down on their public relations offensive. National newspapers ran op-eds about the Golden Wattle as a symbol of unity and identity. School textbooks, botanical gardens, and museum exhibits all quietly aligned with the new nomenclature. By the time international delegates were booking flights to Melbourne, the idea that Acacia belonged to Australia had been domesticated.

Meanwhile, the opposition was busy trying to organize. Botanists from Africa, Asia, and Latin America, many of whom had been blindsided in Vienna, started coordinating letters of protest and formal submissions. The arguments were strong. They pointed out that the original African type species had a documented legacy going back Millennia. They reminded the congress that taxonomic priority, not national symbolism, was the gold standard of scientific naming.

It's important to note that following the Vienna coup, both the IAPT's own post-Congress report and Moore's report warned that this "minority-rule" outcome violated the Code's democratic safeguards and set a dangerous precedent for future nomenclatural disputes [3]. Australian influence within the IBC literally changed the rules for their own benefit. The Australians also anticipated the coming resistance and had their big guns ready for Melbourne.

As the Melbourne Congress approached, a counterattack emerged in the form of academic publications that framed the Vienna outcome as a stabilizing force. Thiele et al. (2011), in a preconference article, warned that reopening the Acacia debate would plunge botanical nomenclature into chaos [4]. These were not passionate manifestos; they were cold, technocratic warnings designed to trigger procedural anxiety. The message was clear: re-litigating the Acacia decision wasn't just wrong, it was dangerous.

In response, in mid-2011, Moore, Murphy, and Weston published a rigorous rebuttal in Taxon (60:3) to Thiele et al.'s pragmatic defense of the 2005 Vienna procedure. Moore's publication was titled "The Acacia Controversy Resulting from Minority Rule at the Vienna Nomenclature Section". Moore's rebuttal carefully documents procedural anomalies at Vienna, including the use of a minority-rule interpretation that allowed the proposal to pass with less than 60 percent support, unprecedented in IBC history [5]. Moore et al. emphasized that key committees endorsed the change, effectively reversing the burden of proof, and assert this undermines democratic safeguards within the Code. They warned that validating the change through a minority decision risks long-term damage to the governance and credibility of botanical nomenclature. Co-authored by eleven botanists from multiple continents (including South Africa, Ethiopia, the UK, Argentina, and the United States) the article reflects global concern rather than a single author critique. The authors call for the erroneous retypification to be separated from the 2006 Vienna Code and revisited formally at the Melbourne 2011 Congress, a move they argue is both democratic and procedural.

This response became the centerpiece of African and Asian botanists' advocacy, reinforcing calls for transparency, majority rule, and equitable representation in future nomenclatural decisions.

THE MELBOURNE CORONATION

When delegates arrived in Melbourne in July 2011, they were met by a city draped in wattle symbolism. Yellow and green banners welcomed attendees, and the local organizing committee provided smooth logistics, guided tours, and impeccable hospitality. The symbolism was subtle but omnipresent. You were not merely attending a botanical congress; you were entering the homeland of the New Acacia.

The African and Latin American delegations arrived in smaller numbers. Many struggled with travel funding, institutional support, or visa access. Several potential attendees from underrepresented nations never made it to Melbourne at all. Those who did make it to Melbourne found themselves on the defensive, outnumbered, and procedurally boxed in. African and Asian delegates felt the outcome was secured even before the vote. A motion to revisit the Vienna decision was introduced but fell flat.

The vote at the 2011 Melbourne International Botanical Congress regarding the adoption of the Vienna Code, and, by extension, the Acacia decision to appropriate Acacia for the Australian wattles, was taken by card and resulted in 373 in favor and 172 opposed, out of a total of 545 ballots cast, yielding a 68% majority endorsement approving Acacia for the Australian wattles [7].

68% of the delegates voted to uphold the 2005 ruling, this time not just winning by technicality, but by overwhelming majority. It was, by any political standard, a rout. The Australian interest group secured their fourth score, and the Acacia heist was complete. At least they believe so.

After Melbourne 2011, Dr. Haripriya Rangan (an ecologist who witnessed the debates) described the sentiment as an expression of "territorial chauvinism", a push for national naming rights that felt deeply colonial [8]. The bitterness of this view illustrates how much more than technical botany was at stake for many. In their seminal study, Kull and Rangan reveal that the debate over the Acacia name was as much about identity and power as it was about taxonomy. Drawing on interviews and firsthand observations at the 2011 Melbourne Congress, they document how Australian delegates mobilized national pride (invoking "golden wattles" as icons of the continent's biodiversity) to frame the name conservation as safeguarding a unique ecological heritage. Australia's message was calibrated and centered on the premise that reversing the decision would not only undo a botanical classification, but it would also insult a nation that had embraced the Acacia name as part of its very identity.

Meanwhile, scientists from Africa and Asia perceived this push as a strategic assertion of botanical "ownership", a neocolonial maneuver that risked erasing centuries of African cultural and medicinal associations with Acacia. The authors argue that this clash of "sentiment" and science crystallized broader tensions between global equity and Western institutional authority. By highlighting moments when delegates referenced national emblems and economic projections alongside botanical arguments, Kull and Rangan demonstrate how nomenclatural decisions can mask political agendas. Their analysis underscores that, beyond morphological characters or DNA sequences, botanical names carry stories of power, place, and belonging. Stories that, in this case, rewrote the legacy of a 2,500-year relationship between people and the African Acacia [6].

The real brilliance of the Melbourne maneuver wasn't the vote; it was the way the Australians managed the entire event. Procedural obstacles were anticipated and neutralized. Dissenting voices were heard but not amplified. No protest made it to the plenary floor. No media scandal erupted and crucially, no new proposal to restore Acacia to its African roots was formally introduced for consideration. The opposition came with a legal case. The Australians brought an entire court system.

Additionally, Australia's ability to send many delegates became a point of contention as Australians were later accused of conning the congress and stacking the vote by flying in more voting participants than African or Latin American countries could afford. The Melbourne Congress (hosted on Australian soil) concluded that any challenge was bound to fail since it guaranteed that many in the global South simply could not afford to come in equal numbers. Indeed, the African side tried to use procedural mechanisms, such as objecting to the adoption of the Vienna Code, to get a redo, but they were outvoted in Melbourne.

In summary, Australian institutions and perhaps government bodies implicitly threw their weight behind retaining Acacia, motivated by national pride (the Golden Wattle emblem), practical concerns (mass renaming costs), and economic interests (forestry and horticulture). The African and international opponents marshaled a broad coalition grounded in principle (upholding nomenclatural rules and historical priority) and regional pride (protecting Africa's symbolic Acacias). The clash was so poignant that it has been cited as an example of how science is not immune to politics and emotion: "science has always been shaped by the social context… this debate about national symbols…caught attention…far beyond the few Acacia specialists" [8]. In the end, the rules of procedure (and effective lobbying) tipped the balance to the Australian side. Even many who accepted the outcome felt uneasy about the optics.

In private conversations, several delegates reportedly expressed discomfort with the optics but justified their votes in the name of "nomenclatural stability". It was a masterstroke of framing. What had begun as a campaign to rename a genus had been recast as a mission to preserve global taxonomic order. Following the Melbourne vote, many African and Asian botanists expressed fatigue with the ongoing dispute. This time no public statement from SANBI was released immediately after the vote. A few scholars vowed to continue using traditional names in regional journals, but the machinery of global taxonomy had moved on.

Is reversal possible? Technically, yes. The International Botanical Congress can revisit any decision at any time. Unfortunately, the principle of nomenclatural stability now weighs against it. Changing thousands of names again would cause global confusion and disrupt regulatory systems. In addition to that, no new congress has signaled interest in reopening the matter. Moreover, generational turnover means fewer taxonomists today feel invested in re-litigating the Acacia debate. The principle of nomenclatural stability, intended to minimize chaos and confusion in science, now militates against change. Efforts to reopen the case would need to demonstrate that the disruption of reversal would be less than the disruption of maintaining the status quo, a bar that grows higher with each passing year.

Moore et al. noted that the case exposed a deeper fault line between scientific process and geopolitical asymmetry. Taxonomy is not immune to nationalism, funding disparities, or symbolic warfare. The Acacia case wasn't a debate; it was a meticulously prepared campaign.

The Australians didn't just win in Melbourne, they closed the case. Or did they?

This is not a mere historical footnote but a textbook case for reversal. Any decision born of clandestine rule-bending, unequal representation, and overt dismissal of governing statutes is "invalid ab initio". The IBC must convene a special session, before the next general vote, to rescind Proposal 1584, reaffirm the genuine supermajority requirement, and restore Acacia to the species that built its legacy. Anything less would be tacit endorsement of institutional chicanery and a dangerous precedent for every future botanical dispute.

Regarding the mechanics, the structural imbalance remains. Wealthier countries like Australia can still dominate international congresses by funding delegate attendance, publishing in top journals, and organizing events.

Whether the world will one day see a reversal, or whether the name Acacia will forever evoke the memory of a heist pulled off with surgical precision, only time will tell. For now, Australia stands as the architect of a rare and lasting victory in the annals of botanical nomenclature. A victory achieved not by chance, but by crooked design.

Now that we have exposed the hidden backdoor and voting details, we will give the actual Australian arguments a chance. In the next chapters we will pretend as if there is no historical, religious, or cultural context. We will pretend that there was no procedural misconduct. We will counter the Australian arguments one by one. Did the outcome reflect justice, or did Australia successfully turn the International Botanical Congress into a Kangaroo Court? Let's step back and take a critical comprehensive look at the arguments from a multi-faceted perspective and determine whether justice, or power, prevailed in this global renaming saga.

References:

1. International Association for Botanical and Mycological Societies. IABMS Constitution, Article 3. https://www.iaptglobal.org/iabms-constitution.
2. Australian Systematic Botany Society. "Melbourne's Successful Bid to Host the XVIII International Botanical Congress in 2011." ASBS Newsletter 124 (September 2005).
3. McNeill, John, Nicholas J. Turland, H. M. Burdet, et al. "Report of the Nomenclature Section of the XVII International Botanical Congress: Vienna, Austria, 12–16 July 2005." Regnum Vegetabile 146 (2006): 58–72.; Moore, Gerry. "The Handling of the Proposal to Conserve the Name Acacia at the 17th International Botanical Congress — An Attempt at Minority Rule." Bothalia 37, no. 1 (May 2007): 109–118.
4. Thiele, Kevin R., Lyn G. Cook, Rod Peakall, and Jeremy J. Bruhl. "Retypification of Acacia Mill.: Stability Versus Majority Rule." Taxon 60, no. 1 (2011): 194–198.
5. Moore, Gerry, Daniel J. Murphy, and Peter H. Weston. "The Acacia Controversy Resulting from Minority Rule at the Vienna Nomenclature Section." Taxon 60, no. 3 (2011): 852–857.
6. International Association for Plant Taxonomy (IAPT). Acacia Beyond Melbourne: Understanding the Decision. IAPT, 2011.
7. McNeill, John, Nicholas J. Turland, Anna M. Monro, and David L. Hawksworth. "Report on Botanical Nomenclature—Melbourne 2011: A Review of the Nomenclature Section at the XVIII International Botanical Congress." PhytoKeys 41 (2014): 1–78.
8. Kull, Christian A., and Haripriya Rangan. "Science, Sentiment and Territorial Chauvinism in the Acacia Name-Change Debate." In Terra Australis 34: Australia's Arid Zone, edited by Chris Clarkson, 205–207. Canberra: ANU Press, 2012.

CHAPTER 8

IBC REASONS FOR ACCEPTING THE 2003 PROPOSAL

The "Report of the Nomenclature Section-Vienna (2005)" makes clear that delegates did not endorse every ancillary argument provided by Australia but instead agreed only with those points directly tied to the Code's criteria for conserving names [1]. In particular, the Report (§ 60.1) records that Proposal 1584 was approved because it:

1. "Best meets the criterion of nomenclatural stability by minimizing the total number of name-changes", under Article 14 of the Vienna Code. In other words, changing the type to an Australian species would force only ca. 171 non-Australian taxa into new genera, whereas retaining the African type would have required nearly 1,000 Australian wattles to be renamed. This arithmetic clearly "best serves the stability of botanical nomenclature" (Report § 60.1). In other words, from the IBC viewpoint, their decision was to achieve the greatest nomenclatural stability by minimizing the number of new combinations worldwide
2. "Maintains continuity of a widely used genus". Since Acacia was already entrenched in hundreds of Australian research papers, forestry regulations, and conservation programs, conserving the name for those species satisfied the Code's injunction to avoid unnecessary disruption of names in common use (Report § 60.1). In other words, from the IBC viewpoint, their decision was also to preserve continuity in a genus already widely used across Australian science and industry.

No other justifications (such as "economic significance", "center of diversity" or "national symbolism") appear in the Report itself. Although those themes were raised in supporting publications (Maslin & Orchard 2003). We will dedicate a chapter of this book to each of these two IBC arguments, as they are of prime importance.

For Argument #1 (to achieve the greatest nomenclatural stability by minimizing the number of new combinations changes worldwide), we will conduct a country-by-country tally effectively demonstrating that far more than "only about 170 non-Australian taxa" would require renaming if Acacia was globally reclassified. The approximate number of 170 taxa requiring updates for the African/Asian thorn tree is extremely understated because the figure does not factor in the varieties and subspecies that will also require updating. We will count all of them and present the actual number for all 113 countries.

Our count will demonstrate the actual work required of 113 countries instead of the work required but Australia alone. Since each country must update every accepted taxon it recognizes, the manual count exposes that the IBC's "minimize name changes" argument grossly underestimates the downstream workload outside Australia.

To counterargue Argument #2 (to preserve continuity in a genus already widely used across Australian science and industry), we will delve and discuss the multi-faceted perspective of this argument. We will also base our counterarguments on the IBC's own code and present how its actions violated them.

References:

1. McNeill, John, Nicholas J. Turland, H. M. Burdet, et al. "Report of the Nomenclature Section of the XVII International Botanical Congress: Vienna, Austria, 12–16 July 2005." Regnum Vegetabile 146 (2006): 58–72.

CHAPTER 9

IBC REASON #1- MINIMIZING DISRUPTIVE NAME CHANGES

The proponents contended that retypification would preserve nomenclatural stability by drastically reducing the number of species needing name changes worldwide. At the time of the proposal, the broad genus Acacia sensu lato comprised roughly 1,350–1,380 species globally. They argued that if the original African type were retained (and the Australian species segregated into a different genus), approximately 1000 Australian species would have to change their name to Racosperma, whereas only about 161 African/Asian species would require renaming to Vachellia [1].

By the Australians' calculation, changing the type to an Australian species would cut the total number of Acacia name changes from an estimated ~1274 down to ~392 – sparing nearly 900 species from nomenclatural disruption. This argument of "weight-of-numbers" was explicitly highlighted. The Committee for Spermatophyta, which reviewed the proposal in 2004, noted that the Australian group's species outnumber the African/Asian species by at least 13:1, a point that "tipped the balance strongly in favor of the proposal" during committee deliberations. Proponents argued that fewer name changes would mean less confusion and labor for herbaria, researchers, ecological databases, laws, and educational materials worldwide.

Counterargument:

To make an honest counterargument, we must first ask ourselves why

IBC REASON #1- MINIMIZING DISRUPTIVE NAME CHANGES

Australia has so much species compared to African/Asian Acacia Species. Why is it that Africa (multiple times the size of Australia) and with many different ecoregions have a significantly smaller number of Acacia species compared to Australia?

Consider todays incorrectly named African/Asian Acacia species. They occupy every major ecoregion: Mediterranean scrub (A. tortilis subsp. tortilis), Sahara dunes (A. tortilis subsp. raddiana), Nile riverbanks (V. nilotica), coastal lagoons near the Red Sea (S. laeta), Indian Ocean lagoons where the fever tree (A. xanthophloea) holds sway, Savanna plains in Kordofan (S. Senegal) and the windswept Cape shores (A. karroo). Yet, across nearly 30 million km² (almost four times Australia's landmass) they total only 379 species.

Conventional intuition attributes Australia's richness to its isolation and age. However, a closer look reveals a different story of taxonomic philosophy. Australia's botanical tradition has, since the days of Baron Sir Ferdinand von Mueller, prized splitting over lumping, elevating every subtle difference (mostly morphological twists of phyllode or shifts in pod shape) to full species status [2]. This juxtaposition defies the classic species–area relationship $S = cA^z$: with a four-fold jump in area a z-value of ≈ 0.25 predicts only a 1.4-fold rise in species [3]. If Australia supports 1,142 wattle species, Africa should harbor roughly 1,600–1,940. Instead, it manages only 379, barely a quarter of the expected total.

Why? Because Australia's splitting culture magnifies nominal diversity. In the mid-1800s, Mueller described almost 800 new plant species, often based on minor morphological distinctions. Differences that, elsewhere, might have been named as varieties [2]. A century later, Leslie Pedley's 1986 revision formalized this ethos, carving Acacia s.l. into three genera and appropriated the 960 Australian forms to Racosperma, almost all at species ranks [4].

IBC REASON #1- MINIMIZING DISRUPTIVE NAME CHANGES

This practice exemplifies what scholars call "biodiversity nationalism", the strategic inflation of species counts to bolster claims of global conservation importance. Hayden documents how Mexican agencies pursued this tactic to attract bioprospecting and conservation funds, while parallel patterns have been observed in Brazil and elsewhere [5]. By contrast, African and Asian botanists treated populations more conservatively, grouping variants under broad species umbrellas and using subspecies or varieties to note local forms.

This philosophical rift had real consequences. In 2005, when the International Botanical Congress redefined Acacia around an Australian type, the decision leaned heavily on Australia's inflated species count. Effectively punishing African and Asian taxonomists for adhering to a different tradition. Yet the "disruptive name changes" Australia lamented were largely of its own making. The fallout extends beyond nomenclature. Taxonomic inflation skews biodiversity metrics, misdirects conservation funds toward hyper-narrow endemics, and fragments scientific communication across "taxonomic dialects". When every minor variant is a species, the signal of true evolutionary lineages is lost in the noise.

So, what's the path forward? Borrowing from integrative taxonomy, we can marry the best of both worlds. We can uphold morphological attention to detail, but demand corroboration from genetic, ecological, and reproductive data before conferring species status. This way, we honor Australia's fine-grained botanical insight without letting splitting philosophy override evolutionary reality.

In the end, it's a matter of intellectual honesty. Australia's prodigious wattle numbers reflect a splitter tradition, not an unparalleled evolutionary explosion. Recognizing this means accepting that the burden of nomenclatural upheaval must rest with the tradition's stewards, not imposed on those who chose a different course.

IBC REASON #1- MINIMIZING DISRUPTIVE NAME CHANGES

I'm sure you are asking yourself, "What difference does it make if the species' counts are inflated due to Australia's taxonomic methodology? Even if the species numbers are not inflated then surely, they would simply have more subspecies and varieties. And those will require updating anyway. Right?" You are absolutely correct.

The African/Asian Species have much less "species" counts compared to Australia's species count but they also have a much higher number of "subspecies" and "Variants" (subspecies and variants are together known as Infraspecies) that would also have to be updated and therefore need to be counted in any "nomenclature stability" comparison. The experts at the IBC surely knew this very well but chose to sideline it. Therefore, to compare apples to apples, the African/Asian species update count would have to reflect the species count in addition to the subspecies and variant counts.

Going back to our counterargument, we must consider all the species and their Infraspecies (variants and subspecies) which will also have to be updated within the local databases of all the countries they are native to. Since the vast majority of Australin Acacias are in Australia, there is only one system to update. For the original Acacia thorn tree (now incorrectly called Vachellia), however, that is not the case, as the Acacia thorn tree is native to more than 113 countries and each of these countries would be required to update their "Vachellia" species in addition to their Infraspecies. As you can see, the numerical argument quickly becomes irrelevant when you understand that the IBC and the Australian lobby purposefully left out the Infraspecies level updates from their comparative equation. They also conveniently left out the number of African and Asian countries that will bear that expense.

Unfortunately, we cannot place a numerical weight value to the efforts of 113+ countries that will incur the burdensome updates under Australia's proposal 1584. However, we can count how many total update entries are required in all these 113+ countries.

First, I had to make sure that the Vachellia Species numbers today are the same as they were in 2004 when proposal 1584 was approved. The "thorn-tree" block that still bore the name Acacia at the Vienna Congress in 2005 comprised roughly 165 species "about 161 names of subg. Acacia" yet to be transferred, according to the official Committee report [6] and reiterated by Luckow et al. (only four Vachellia combinations then existed) [7]. Two decades later, authoritative checklists give 158–164 accepted Vachellia species. "Plants of the World" Online lists 158, while "WorldWideWattle" tallies 164 [8], showing that the 2005 residue and today's total differ by only a handful.

It's also important to note that since the 2005 retypification of Acacia, only about 4–6 new Vachellia species have been formally described or taxonomically revised in the scientific literature. This means that "Vachellia" species numbers have only been increased by 4-6 [9]. Today's list (~160 spp.) is essentially the same cohort that delegates debated in 2005.

Finally, to come up with the total number of update entries for the now Vachellia genus, I manually ran an advanced search in the online Australian database WorldWideWattle for each of the 113+ countries where today's Vachellia are native [10]. For each country, I set:

- Genus to Vachellia
- Name Status to "Accepted" (excluding synonyms or outdated names like Mimosa)
- Provenance to "Native"
- Country to the specific nation

IBC REASON #1- MINIMIZING DISRUPTIVE NAME CHANGES

By recording the "× records" count for each search, I captured the exact number of currently accepted taxa (both species and infraspecies) that will require a name change if those genera are reclassified according to Proposal 1584. The sum of these country-level totals gives the total database entries each nation must update. This approach ensures an "apples-to-apples" comparison of "update entries". Just as Australia would have to update 1,202 Acacia entries (1,075 species + 127 infraspecies) if its wattles moved to Racosperma, every respective country (113+ of them) with Vachellia must update their own accepted taxa. Because each native taxon appears once per country, the grand total reflects the true scope of work. This is the only way to avoid unfairly comparing Australia's count to only 161 global species since infraspecies would also need updating. I recorded only accepted names, ensuring accuracy and consistency.

IBC REASON #1 - MINIMIZING DISRUPTIVE NAME CHANGES

Country	Vachellia Species	Vachellia Infra
Afghanistan	1	0
Algeria	3	4
Angola	12	9
Argentina	7	1
Australia	9	2
Bahamas	5	0
Bangladesh	2	1
Barbados	2	0
Belize	9	1
Benin	2	0
Bolivia	6	1
Botswana	17	12
Brazil	3	0
British Virgin Islands	2	0
Burkina Faso	4	1
Cambodia	1	0
Cameroon	3	0
Cayman Islands	1	0
Central African Republic	2	2
Chad	6	5
Chile	5	0
Columbia	4	1
Costa Rica	8	0
Cuba	12	0
Democratic Republic of Congo	9	7
Djibouti	6	3
Dominica	1	0
Dominican Republic	8	0
East Timor	2	0
Ecuador	8	3
Egypt	4	4
El Salvador	6	2
Ethiopia	27	20
French Guiana	2	0
Ghana	5	2
Grenada	2	0
Guadeloupe	3	0
Guatemala	12	2
Guinea	1	0
Guinea Bissau	2	1
Guyana	2	0
Haiti	4	0
Hispaniola	1	0
Honduras	9	3
India	12	6
Indonesia	2	0
Iran	4	4
Iraq	2	4
Israel	2	2
Ivory Coast	3	0
Jamaica	3	0
Kenya	24	20
Kuwait	1	2
Laos	2	0
Lesotho	1	0
Lesser Antilles	1	0
Libya	2	3

IBC REASON #1 - MINIMIZING DISRUPTIVE NAME CHANGES

Country	Vachellia Species	Vachellia Infra
Madagascar	3	0
Malawi	10	0
Mali	7	6
Martinique	2	0
Mauritania	2	0
Mexico	33	6
Montserrat	2	0
Morocco	1	0
Mozambique	18	8
Myanmar	6	1
Namibia	12	12
Nepal	1	1
Netherlands Antilles	3	0
Nicaragua	7	2
Niger	6	3
Nigeria	7	9
North Yemen	13	10
Oman	6	8
Pakistan	5	7
Palestine	1	2
Panama	6	0
Paraguay	5	1
Peru	5	3
Puerto Rico	3	0
Qatar	4	2
Rwanda	5	3
Saudi Arabia	11	14
Senegal	4	7
Socotra	2	0
Somalia	20	17
South Africa	29	10
South Yemen	11	8
Sri Lanka	5	0
St Kitts-Nevis	2	0
St Lucia	1	0
St Martin	2	0
St Vincent	1	0
Sudan	16	22
Suriname	1	0
Swaziland	11	4
Syria	2	1
Tanzania	29	22
Thailand	4	0
The Gambia	1	1
Togo	5	1
Turks & Caicos	1	0
UAE	1	1
Uganda	15	11
Uruguay	2	0
US Virgin Islands	1	0
USA	11	2
Venezuela	5	2
Vietnam	3	0
Western Sahara	2	1
Zambia	12	8
Zimbabwe	23	14

IBC REASON #1- MINIMIZING DISRUPTIVE NAME CHANGES

Before I present the final infraspecies number, I must make a disclaimer that the infraspecies numbers i'm presenting within all these countries are significantly understated. Even when regional botanists recognize distinct subspecies or varieties. Many of these infraspecific names never reach international databases simply because they remain "hidden" in local floras, unpublished theses, or obscure journal articles. Limited funding and staffing mean that many herbaria lack the resources to prepare full, peer-reviewed descriptions in English language outlets. Additionally, publications in national or regional bulletins rarely triggers indexing by global repositories like IPNI or GBIF.

Language barriers and paywalled local journals further block data flow, so even well-documented variants in Central African or Amazonian monographs fail to appear in widely used portals. Without formal publication under the International Code of Nomenclature (including type citations, Latin or English diagnoses, and registration) these names cannot be harvested by "Plants of the World Online" or "WorldWideWattle". As a result, genuine infraspecific diversity remains under-represented worldwide until concerted efforts convert local knowledge into internationally accessible, databased records. That said, according to the most comprehensive dataset available, there are currently:

- 113 countries that contain Vachellia species and its infraspecific taxa, with a total of 1,063 name-bearing taxa.

This results in a combined global name-change burden of 1,063 operations for 113 countries outside Australia per Proposal 1584. These operations are not theoretical, they involve real updates to national biodiversity databases, flora catalogs, herbarium records, regulatory documents, environmental policies, educational materials, and legal designations. Each of these 113 countries must undergo these operations independently, using separate administrative and technical systems. By contrast, Australia has

IBC REASON #1 - MINIMIZING DISRUPTIVE NAME CHANGES

approximately 1,212 Acacia taxa (species, subspecies, and hybrids) that would have required renaming if the name Acacia had not been conserved for Australian taxa. Importantly, all these edits would occur in a single country, using a centralized infrastructure.

In numerical terms:
- Global workload for African/Asian origin taxa (Vachellia): 1,063 name-change operations across 113+ national systems.
- Australian workload if retypified: 1,212 operations in 1 national centralized system.
- Resulting difference: The real difference is only 149 additional name change operations that could have been much easier to implement in just one country under one system. Not the "13:1 difference" directly quoted verbatim from the committee for Spermatophyta [6].

This calculation dismantles the central rationale presented by Australian delegates and finally accepted at the 2005 IBC in Vienna. The claim of "minimizing disruption" was narrowly framed around species count rather than the distribution and complexity of that disruption. While it is true that Australia has more endemic Acacia species than any other country, these species are restricted to a single, coordinated taxonomic infrastructure. In contrast, African/Asian Acacia species are spread across nearly 113 countries, including many with limited botanical capacity and decentralized bureaucracies.

Even from a purely numerical standpoint. Setting aside cultural, economic, and infrastructural concerns, the facts are damning. A total of 1,063 discrete retypifications across 113 countries represents a disproportionate taxonomic cost to the global south, all to spare a single rich country from renaming 1,212 species through their computerized system. The nomenclatural decision of 2005 thus reflects not a neutral, disruption-minimizing compromise, but a geopolitically lopsided resolution. It

IBC REASON #1- MINIMIZING DISRUPTIVE NAME CHANGES

exported the overwhelming burden of renaming onto the Global South in favor of protecting convenience within Australia.

This analysis also anticipates and neutralizes the potential rebuttal that species count is the sole metric that matters. If name-change operations were simply about numbers, one might initially think 1,212 edits in Australia exceed any single-country burden. However, disruption is not just about how many species are renamed; it is about how widely those names are entrenched in one's culture and how many separate systems must be altered to accommodate the changes.

In practice, the process of reclassifying a species is not simply a find-and-replace operation. It involves institutional vetting, expert consultation, database modification, and often public-facing educational campaigns. In multilingual nations or those with significant reliance on indigenous knowledge systems, the challenge becomes even more complex. Moreover, many of these Acacias are culturally or economically significant. In Sudan, Senegal, and Kenya, for example, Acacia Seyal is a vital source of Acacia gum, exported globally for use in food and pharmaceuticals. Renaming these species across administrative, commercial, and ecological registers involves costly and time-consuming transitions.

Had the IBC allowed Australia to reclassify its own 1,212 Acacia taxa into an alternate genus, all edits would have been managed within a single digital infrastructure, under one taxonomic community, one language, and one government framework. Instead, the rest of the world must now execute almost the same amount of name changes, across dozens of legal and linguistic environments with far fewer institutional resources.

That said, at the heart of botanical nomenclature lies the principle of priority. The idea that the earliest validly published name for a genus or species should be preserved to ensure historical continuity and scientific consistency. In the case of Acacia, this principle should have safeguarded the African lineage, where the name was first typified and historically used for species like Acacia nilotica and Acacia Seyal. Yet, in a controversial decision, the International Botanical Congress (IBC) in 2005 voted to conserve Acacia for the Australian wattles, citing "minimization of nomenclatural disruption" as justification.

This rationale, however, loses force when we scrutinize the actual numbers. Our updated calculation shows that 1,063 African/Asian name-bearing taxa (species plus infraspecific ranks) still required reclassification after 2003, compared to 1,212 taxa in Australia that would have needed renaming if Acacia had remained with its original type. The difference (just 149 taxa) is remarkably narrow, especially considering the global fragmentation of African and Asian taxa across 113+ national systems, each of which must carry out its own updates.

When the burden is viewed not as a raw count of names but as the cumulative impact across global jurisdictions, it becomes clear that the Australian decision externalized the disruption rather than reduce it. Considering such a close numerical margin and given the historical and geographic origin of the name Acacia, the IBC should have adhered to its own mandate of priority. Upholding Acacia for the African lineage would not only have honored nomenclatural integrity but also distributed the update burden more equitably, logically and efficiently.

In conclusion, when one incorporates detailed counts of species and infraspecific taxa, the data show unambiguously that the IBC decision to reassign Acacia to the Australian clade did not minimize disruption. It externalized and magnified it, imposing an incalculable larger global burden, distributed across countries least equipped to absorb it.

References

1. Anthony E. Orchard and Bruce R. Maslin, "Proposal to Conserve the Name Acacia (Leguminosae: Mimosoideae) with a Conserved Type," Taxon 52, no. 2 (2003): 362–63.
2. Ferdinand von Mueller, Fragmenta Phytographiae Australiae, 12 vols. (Melbourne: Government Printer, 1862–81); Bruce R. Maslin, "Taxonomic 'Splitting' in Acacia: Historical Trends and Their Consequences," Australian Systematic Botany Society Newsletter 95 (1998).
3. "Species–Area Relationship," ScienceDirect Topics, Elsevier, https://www.sciencedirect.com/topics/earth-and-planetary-sciences/species-area-relationship.
4. Leslie Pedley, "A Synopsis of Racosperma C. Mart. (Leguminosae: Mimosoideae) in Australia," Austrobaileya 6, no. 3 (1986): 445–96.
5. Cori Hayden, When Nature Goes Public: The Making and Unmaking of Bioprospecting in Mexico (Princeton, NJ: Princeton University Press, 2003).
6. R. K. Brummitt, "Report of the Committee for Spermatophyta: 55. Proposal 1584 on Acacia," Taxon 53, no. 3 (2004): 826–29.
7. Melissa Luckow et al., "Acacia: Consequences of Typification Choices," Taxon 54, no. 2 (2005): 513–19.
8. Royal Botanic Gardens, Kew, "Vachellia," Plants of the World Online, https://powo.science.kew.org; Bruce R. Maslin, "Species Numbers," WorldWideWattle, Western Australian Herbarium, https://worldwidewattle.com.
9. Bernard Kyalangalilwa et al., "Four New East-African Vachellia Species (Fabaceae: Mimosoideae)," Bothalia 43 (2013): 15–24; Albert Chikuni et al., "A Revision of the Vachellia seyal Complex (Fabaceae: Mimosoideae)," Kew Bulletin 67 (2012): 235–44.
10. WorldWideWattle Database, Western Australian Herbarium, https://worldwidewattle.com.

CHAPTER 10

IBC REASON #2 – BRAND IMPORTANCE

The IBC's second reason for accepting the Australian proposal of 2003 was because, "Australian botany, forestry, agriculture, and related industries rely heavily on the name Acacia, so preserving it avoids confusion and economic impact in Australia". This second rationale, preserving continuity in Australia, seemed straightforward and defensible. However, any astute professional will caution as this rationale is deceptively shortsighted. It focused on immediate, visible disruptions in a single wealthy country while ignoring the slow, compounding burden imposed on dozens of poorer nations. To invoke continuity while discarding equity is not scientific governance. Rather it's a privilege of systemic convenience for the wealthy at the direct cost of the structurally disadvantaged.

Let us begin by unpacking the premise. The IBC claimed that Acacia is deeply embedded in Australian scientific and industrial infrastructure. This includes botany textbooks, forestry manuals, ecological research, and commercial branding. The golden wattle (Acacia pycnantha) stands as a symbol of national identity, but symbols are not scientific arguments. They are public relations tools, and privileging symbols over data and nomenclatural consistency sets a dangerous precedent in global taxonomy.

Australia's Acacia species (in 2001 they 1,020 taxa including subspecies and hybrids and now 1,212 in total) do indeed populate a well-documented national flora [1]. This reality, however, masks the fact that renaming those species under Racosperma would have required a one-time, centralized update. The Australian National Herbarium, APNI (Australian Plant Name

Index), and other digital platforms could implement this in batch processes with relatively higher coordination, minimal cost, unified language and infrastructure [2].

Contrast this with the African and Asian reality. The IBC's continuity rationale claimed to protect Australia from disruption. The reality is that while renaming 1,212 Australian taxa in a single jurisdiction may seem significant, it pales when compared to the real, distributed consequences of the IBC's decision. Almost the same number of name-bearing taxa still required nomenclatural revision due to IBC's decision, but the difference is these were updates across more than 113 independent systems. These include national databases, regional floras, agricultural registries, pharmaceutical documents, and legal export codes.

One cannot overstate the asymmetry of this burden. Australia, a country with a GDP over $1.3 trillion, was shielded from a controlled, centralized taxonomic update. Meanwhile, dozens of global south nations, countries with strained taxonomic funding and weaker digital infrastructure, were forced to wrestle with renaming species critical to trade, agriculture, and environmental science. This is especially true for commercial Acacia gum from Acacia seyal. As one of Africa's largest export commodities, Acacia seyal was part of the 170-species contingent still bearing the Acacia name. In nations like Sudan, Chad, and Nigeria, A. seyal is enshrined in phytosanitary documents, export certificates, chemical registries, and scientific reports. Renaming it means reeducating customs officials, rewriting trade agreements, and synchronizing databases. This level of effort requires coordination that many of these countries are ill-resourced to achieve. Yet they were asked to shoulder this burden so that Australia could maintain lexical familiarity in herbarium labels.

Let's examine this from a fair perspective. The IBC's decision removed friction from the path of the powerful while transferring uncertainty to

IBC REASON #2 – BRAND IMPORTANCE

those least equipped to manage it. This is the very mechanism that produces fragility in global systems. The appearance of continuity for one becomes the generator of entropy for many.

Keep in mind taxonomy is a science defined by regular upheaval. Names change as new data emerges. Entire genera are split or lumped as cladistics, DNA sequencing, and phylogeographic studies evolve. Why should Australia be granted an exemption from this universal process? The IBC's rationale assumed Australian continuity was more critical than global consistency. However, nomenclature is a shared language, not a national asset. If a name change in one country inflicts more total disorder abroad than at home, then preserving it is not stability, it is parochialism masquerading as prudence.

Moreover, the argument hinges on the unquantified assumption that Australian industries, regulators, and scientists would suffer economic harm from the renaming. This claim lacks rigorous evidence. There are countless examples of scientific renaming that did not collapse industries. The reclassification of Cassia into Senna, the reshuffling of Calathea into Goeppertia, or even the splitting of Lilium in horticulture. All proceeded with adaptation, not paralysis. Industries adapt, labels get updated, and science continues. The wattle situation was no different, except it offered Australia a diplomatic chance to support global scientific equity. Instead, it chose lexical comfort.

To argue that Acacia pycnantha must remain Acacia because it is on the national crest is not a scientific claim. It is an aesthetic preference elevated to taxonomic law. That elevation came at the expense of more than one hundred countries that now must operate under disrupted, bifurcated, and inconsistent naming systems. This imbalance is further compounded by the global demand for transparency and accuracy in trade. Exporting Acacia seyal products now involves resolving discrepancies between what local floras call it and what international buyers expect. If an import inspector in Germany looks for Acacia seyal and instead sees Vachellia seyal on the certificate, what delays or confusion ensue? If Sudan, Ethiopia, or Nigeria can't afford to update all relevant documentation, do their exports become legally questionable or noncompliant?

This is no longer about names. It is about access, credibility, and sovereignty in international markets. The IBC's preservation of Australian continuity created dissonance in the African/Asian scientific, economic and legal continuity. The justification sidelined the global in favor of the national, and in doing so, reified a colonial asymmetry that scientific institutions have a duty to dismantle, not reinforce. I therefore present five counterarguments to this one argument even though only one of these would suffice.

1. Temporary Adjustment vs. Long-term Stability

The first error in the IBC's reasoning is what we might call a failure of "temporal asymmetry". Imagine two paths: 1. The first involves a short-term centralized renaming effort within Australia, creating immediate disruption but leading to long-term global standardization. 2. The other path preserves Australia's peace at the cost of decades of fragmented confusion for 113+ countries. The IBC chose the second path, which was superficially stable but globally corrosive. By letting Australia avoid pain, it consigned the rest of the world to permanent disorder. That's not continuity. That's amortized entropy.

Australia's internal systems, digitized and coordinated, could have implemented a change in a mere matter of months. Compare that to the distributed network of global south nations. These regions span dozens of languages, systems, and legal standards. Updating the name of Acacia seyal isn't just about taxonomic records, it cascades into every downstream function: from warehouse labeling in Dakar, to customs processing in Djibouti, to pharmaceutical coding in France.

The IBC chose short-term comfort for one country over long-term clarity for the entire planet. This is a classic case of "intervention bias", mistaking avoidance of discomfort for optimization. The IBC argued for minimizing confusion. However, by choosing to preserve the Australian type, it maximized the geographic scope, complexity, and inequity of the disruption. As statisticians will say: it created an "inverse Pareto fragility", where the smallest group absorbed the least pain, and the largest group paid the price for the small group's stability.

2. Digital and Institutional Capacity in Australia

Australia is not Sudan. It's not Niger. It's not Chad. This is not an insult; it is a logistical reality. Australia boasts one of the most comprehensive botanical databases in the world. The Australian Plant Name Index (APNI), the Australian National Herbarium, and national agencies like CSIRO possess the digital backbone, institutional funding, and technical personnel to implement nomenclatural shifts with stunning speed.

By contrast, many African nations rely on partially digitized, donor-dependent infrastructure. A name change doesn't just require a word doc update, it demands institutional retraining, system-wide coordination, and legal re-issuance of export documentation. What the IBC did was give the easiest burden to the party most capable of bearing a hard one, and the hardest burden to the parties least able to adapt. That is injustice engineered through privilege.

3. Ability to Adapt Quickly and Efficiently

Industries in Australia are no strangers to change. From changes in water regulation to carbon pricing schemes to labeling standards for produce.

Australian enterprises routinely adjust to new regulatory frameworks with minimal lag time. Why would botanical names be any different?

The Acacia reclassification would have affected sectors like forestry, academia, and nursery businesses. Yes, that impact would have been real, but it would also have been ephemeral. One-time changes to catalogs, brochures, signage, and databases would have caused a ripple, not a rupture.

Contrast that with the non-adaptive environments in which Acacia seyal exists. Its name appears on sacks of gum in Arabic, French, Amharic, Hausa, and Swahili. Its documentation spans customs reports, border-control logs, NGO nutrition protocols, country specific regulatory bodies and UN import-export databases. Moreover, the IBC chose to protect Australian nurseries over African economies. That's not continuity, that's strategic insulation masquerading as diplomacy.

IBC's decision should not have been about maintaining the convenience of a single nation. It should have been about absorbing shocks with minimal cost to the whole system. The IBC chose the most damaging outcome possible which is localized comfort in exchange for global incoherence.

4. International Trade Implications: When Names Control Markets

The name Acacia seyal is not merely botanical nomenclature. It is an economic passport. As a cornerstone of the global Acacia gum industry, it defines tariff codes, phytosanitary certificates, and laboratory assays across dozens of countries. Changing that name in fragmented, under-resourced nations introduced confusion in the bureaucratic setting which escalated into a catastrophic trade rejection of the name change policy itself. To this day, neither the FDA, nor the EFSA nor any of the major market's regulatory bodies accepted this name change. We will discuss this more in the following chapters.

5. Global Botanical Communication and Clarity

Clarity in nomenclature serves the transmission of knowledge. Acacia in Africa meant something ancient, trade-rooted, and pharmacologically vital. When that meaning is diluted to accommodate Australian preferences, the global dialogue becomes incoherent. What once was a shared term across phytochemistry, anthropology, trade, and conservation is now a source of confusion. When one name must simultaneously signify both A. pycnantha and the African thorn tree of biblical fame, clarity is lost.

Scientific naming should clarify, not confuse. The IBC's decision turned a once-coherent taxonomic map into a politically distorted diagram, one where linguistic cohesion in Canberra caused dissonance in Khartoum.

Conclusion: Privilege as Policy

In conclusion, the IBC's second rationale for retaining Acacia to preserve Australian nomenclatural continuity reveals itself, under scrutiny, as a rationalization of privilege. It presumed fragility where there was resilience (Australia), and resilience where there was fragility (Africa). It elevated narrative over data, lobbying over logic, and symbolism over science.

To defend this decision is to defend a taxonomy governed not by principles of clarity, equity, or scientific integrity. It is a defense of the loudest voice in the room. The IBC decision was a hijacked process under the illusion of stability.

References:

1. Bruce R. Maslin and Phillip G. Kodela, "Introduction to Acacia," in Flora of Australia, vol. 11A, Fabaceae (Mimosoideae)—Acacia part 1, ed. A. E. Orchard and A. J. G. Wilson (Melbourne: ABRS / CSIRO Publishing, 2001), xiii–xxxi.
2. W. R. Barker, A. E. Orchard, and P. G. Kodela, "The Australian Plant Name Index (APNI): A Tool for Managing Bulk Nomenclatural Changes in the National Flora," Nuytsia 13, no. 3 (2000): 239–52.

CHAPTER 11

OTHER AUSTRALIAN LOBBY ARGUMENTS

In this chapter we will revisit the many other Australian arguments in the 2003 proposal 1584. We will also dedicate the next few chapters to debunk larger arguments in separate chapters. Finally, we will switch gears in the final chapters to make our own arguments that challenge Australia's quest for privileged treatment.

Other Australian Arguments #1 – Nomenclatural Mechanics and Gender of New Names:

Orchard and Maslin also raised a technical point. If the Australian species were renamed under the genus Racosperma (a previously proposed genus for Australian wattles), many specific epithets would need minor spelling alterations due to Latin grammar. Racosperma is neuter in gender, whereas Acacia is feminine; thus, for example, Acacia pycnantha would become Racosperma pycnanthum, A. aneura would become Racosperma aneurum, and so on. The Australians argued that such changes (affecting hundreds of species names) would add further complication, requiring edits to labels, databases, and legal lists where the species name appears. By conserving Acacia for the Australian group, these spelling modifications (and potential confusion over whether two names are in fact the same species) could be largely avoided [1].

Our counter argument is that Latin gender endings are the weakest pillar on which to rest a global taxonomic upheaval. Changing pycnantha to pycnanthum or aneura to aneurum is a purely orthographic adjustment, involving no new types, no new combinations, and merely automatic agreement with a neuter genus. Modern databases already manage such routine spelling shifts: APNI, GBIF, and IPNI store canonical stems and generate gender-matched endings on the fly. A batch script could propagate the revised epithet across Australian herbaria, seed banks, and legislation in a single update cycle. Herbarium label software does it daily for corrections in author citation or rank. In contrast, conserving Acacia for Australia forced 1,000+ hard-code changes in more than 113 countries (each in separate languages, statutes, and supply-chain documents) merely to spare Australian curators a find-and-replace operation.

Furthermore, orthographic tweaks are common currency in botany. The transfer of many Calathea species to neuter Goeppertia required a-to-um conversions that nurseries and databases accepted overnight. No industries collapsed, no legal chaos ensued. Australia's claim that such trivial edits constitute prohibitive "complication" thus rings hollow.

More importantly, Latin endings are invisible to most users and will not diminish a consumer's understanding of the product. Ecologists, policymakers, and exporters care which taxon a name points to, not whether its suffix is –a or –um. The genuine confusion now arises in Africa and Asia, where legally entrenched commodities like gum Acacia (Acacia seyal) must navigate dual nomenclatures because Australia declined a minor orthographic nuisance. Elevating spelling convenience for one wealthy nation over global nomenclatural equity subverts both the spirit and the letter of the Code, which prioritizes historical continuity and universal clarity above parochial comfort.

Other Australian Argument #2 – Cultural Symbolism:

Wattles (the golden wattle, Acacia pycnantha) are Australia's national floral emblem; retaining the name underlined national identity [2]. This argument has been debunked in the first chapters of this book. Australia's move to appropriate the name Acacia purely on recent nationalist grounds represents more than nomenclatural theft; it amounts to historical erasure and cultural injustice akin to symbolic genocide. It dismisses thousands of years of documented human heritage spanning diverse cultures, religions, industries, and traditions. Retaining the name Acacia exclusively for Australian wattles trivializes profound global symbolism and disregards the collective human legacy inherent in these ancient African and Asian species, making the decision fundamentally unjust and historically indefensible.

Other Australian Argument #3 – Clarity of Australian Phyllodinous Clade Monophyly (Phylogenetic Justification):

Molecular and morphological studies have indicated that the Australian phyllodinous species forms a cohesive, monophyletic group. Although less emphasized, this argument bolstered the notion that retypification aligned with a natural phylogenetic delimitation of Acacia sensu stricto [3]. When people following this IBC sponsored theft claimed that Australia changed how the IBC operates, they did not exaggerate. Monophyly alone is not a passport to nomenclatural ownership. Yes, molecular and morphological data confirm that Australia's phyllodinous wattles form a tidy clade, but so do the African/Asian Thorn tree lineages. Taxonomic practice repeatedly accommodates multiple monophyletic genera that share a historically established name by assigning new names to off-shoot clades, not by confiscating the original label. When Cassia was split into the clearly monophyletic Senna and Chamaecrista, the ancestral name was not reassigned to the largest clade; it stayed with the type species, safeguarding historical continuity. The same principle should have protected Acacia nilotica.

Moreover, monophyly can be honored without uprooting two millennia of cultural, economic, and legal usage tied to the African species. Pedley's 1986 solution (retaining Acacia for the African type and renaming the Australian clade Racosperma) respected both evolutionary relationships and nomenclatural stability. The IBC instead privileged aesthetic neatness for one wealthy nation over global coherence, creating a precedent that phylogenetic purity overrides priority, cultural heritage and equitable burden-sharing. By contrast, conserving Acacia for its original African lineage would have required exactly one taxonomic adjustment (Australia accepting Racosperma) and left the rest of the world's nomenclature intact.

Finally, invoking monophyly ignores the Code's hierarchy of values which is priority first, stability second and phylogenetic elegance only when it does not upend both. The 2005 vote inverted that order without batting an eye. True scientific stewardship balances evolutionary insight with historical responsibility and global equity. It does not award an ancestral name to the loudest contemporary majority.

Australian Argument #4 – Economic & Ecological Significance of Australian Wattles:

Australian wattles underpin critical industries (timber, tannin, restoration plantings) and play keystone roles in arid and semi-arid ecosystems. Retaining Acacia for these species preserved consistency in scientific, commercial, and land-management literature therefore avoiding costly confusion in sectors reliant on stable nomenclature (e.g., revegetation projects, seed-certification schemes) [4].

This argument deserves two separate chapters. Let's take a closer look at the economic and ecological relevance of Australian wattles compared the African/Asian species.

References:

1. Maslin, Bruce R., and Anthony E. Orchard. "Proposal to Conserve the Name Acacia (Leguminosae: Mimosoideae) with a Conserved Type." Taxon 52, no. 2 (2003): 362–363.
2. Maslin, Bruce R. Wattle: The Acacia of Australia. Nedlands, WA: Department of Conservation and Land Management, 2001.
3. Miller, Joseph T., and Randall J. Bayer. "Molecular Phylogenetics of Acacia (Fabaceae: Mimosoideae) Based on Chloroplast matK and Nuclear ITS Sequences." Plant Systematics and Evolution 239, nos. 3–4 (2003): 217–232.
4. Maslin, Bruce R., Joseph T. Miller, and David S. Seigler. "Overview of the Generic Status of Acacia (Leguminosae: Mimosoideae)." Australian Systematic Botany 16, no. 1 (2003): 1–18.

CHAPTER 12

ECONOMIC SIGNIFICANCE COMPARED

If you really want to understand which kind of Acacia "rules" the modern economy, don't start in a lab or a botanical archive. Start on the floor of a warehouse, or in the procurement office of a global manufacturer. Better yet, start in the fluorescent lit rooms of Nestlé or Johnson & Johnson, where product decisions are made in milliseconds and measured in SKUs, not species.In today's economy, botanical relevance isn't judged by Latin names or tree counts. It's judged by market saturation. It's judged by how many things (products, codes, inventory units) require that substance. By that measure, African/Asian Acacias are not just important. They're dominant. Overwhelmingly so.

Let's shift our lens. Let's not count trees and instead count SKUs. In the language of commerce, a SKU isn't just a barcode. It's a vote. A presence. A signal that this ingredient is embedded in the machinery of global trade. And when you scan that machinery, what you find is that Acacia gum (from African/Asian Acacia species) is everywhere.In soft drinks, gummies, baby formula, mascara, marshmallows, syrups, fiber powders, lozenges, and pharmaceuticals. It's one of the rare eco-friendly and all-natural bloodstreams of the global economy.

Tens of thousands of SKUs (from Walmart, Amazon, Carrefour and Alibaba) contain African Acacia derivatives. The ingredient is labeled variously: "acacia gum", "gum acacia", "acacia fiber", "gum arabic" and more simply as "INS 414" or "E414." Whatever the label, the presence is staggering. You'll find these SKUs in:

ECONOMIC SIGNIFICANCE COMPARED

- Every major soft drink brand: Coca-Cola, Pepsi etc.
- Confectionery: M&Ms, Licorice, marshmallows, gummies and gum drops.
- Pharma: Tablets, syrups, sprays, supplements, Nano carriers.
- Baby and medical nutrition: Protein powders, baby formulas, nutrition bars etc.
- Functional food and prebiotic drinks.
- Bakery glazes, syrups, icing.
- Cosmetics and personal care: including lotions, mascaras, and powders.
- Industrial applications: adhesives, inks, and textiles.

Acacia gum isn't just "in" these categories, it's integral. It's the literal glue that holds the product's texture, stability, shelf life, and appearance together. In food and beverages, it suspends flavors and colors. In candy and baked goods, it gives chew, shine, and consistency. In food products it's the ultimate prebiotic. In pharma, it delivers actives with precision, binding everything from cough drops to probiotics.

Pharmacopoeias around the world (the U.S., European, and Japanese) specify African/Asian Acacia gum in product after product. These are SKUs that pass regulatory scrutiny and hit store shelves across every market. What about the Australian "wattle"? Entirely missing. Barely any pharmaceutical or over-the-counter product in the U.S., EU, or China lists Australian acacia as a recognized excipient.

In the booming industry of gut health, prebiotic supplements, and "clean-label" nutrition; African/Asian Acacia gum is the kingpin. Thousands of new SKUs in North America, Europe, and the Gulf markets now wear the badge "acacia fiber" or "acacia gum prebiotic", nearly all sourced from Sudan, Chad, or Nigeria [1]. This is where the future of functional food is being written. The Australian acacia derived food products are almost exclusively local to Australia with minimal SKUs.

We can estimate the number of SKUs relying on African/Asian Acacia Gum in North America, EU and Japan at the tens of thousands and possibly hundreds of thousands. These do not include the Acacia Gum reliant SKUs of China, India, and Latin America which import a big proportion of Acacia gum sales worldwide. So, when someone says "Acacia," ask them to show you the SKUs because the species that feeds, glazes, binds, emulsifies, coats, thickens, heals and stabilizes through hundreds of thousands of global products isn't Australian. Taxonomic politics and lobbying will never change that.

The Australian Acacia Economic Relevance

To understand the global impact of a botanical species you start with the shelf, the supply chain, and the storefront. In that story (the story of Australian Acacia) you quickly find yourself lost. While Africa's Senegalia and Acacia Seyal flood the world with fibers, stabilizers, emulsifiers, bulking agents, Nano carriers, and encapsulations; Australian wattles show up as boutique cameos. Beautiful but economically not as relevant.

Start with the food and beverage industry. Australian Acacia-based SKUs are negligible. Most are built around wattle seeds which are earthy, protein-rich, and undeniably regional. You'll find them in local Australian herbal lozenges, polyphenol capsules, and as minor colorants or clarifiers in wine. You will also find them in limited edition ice creams, Gin infusions, and Café pastries. These aren't core products with global scale. They're novelties with small geographic footprints.

Now step into the leather industry. Here Australia's Acacia mearnsii (the black wattle) does make a big dent. Not in its homeland, but abroad. Plantations across Brazil, South Africa, China, and Vietnam strip its bark to make mimosa extract: a vegetable tannin prized for chrome-free leather production [2]. Analysts valued the entire tannin market (mimosa, quebracho, chestnut, tara, oenological variants) at $1.8 to $2.0 billion in 2022 [3]. Acacia mearnsii plantations in Brazil, South Africa, China, and Vietnam are globally important suppliers of mimosa (vegetable) tannin for leather production. More importantly, how many SKUs would change if this tannin were renamed tomorrow? Very few. The impact isn't in product codes; it's in industrial chemistry. These tannin products are economically quiet. unbranded and invisible to consumers [4].

Moving on to timber. The visual catalogues (TimberSearch ANZ, Wood-Database) list a modest assortment of "Black Wattle" codes: dressed boards, turning blanks, some flooring [5]. A few dozen SKUs proving this is not an industry driver.

Then there's wax. A newcomer to the scene, Acacia decurrens flower wax (formally listed as Acacia Decurrens Flower Cera) is gaining traction in cosmetics. Used for film-forming, structuring, and solid textures [6]. The wax is a nice idea and a natural option, but yields are low and not commercially significant. Other Australian Acacia species (such as A. longifolia) have been chemically evaluated for leaf cuticular waxes and phenolic compounds, with studies highlighting their potential utility in biodegradable cosmetics and timber surface coatings [7]. All sounds promising in local headlines, but the numbers reveal the limited potential. With yields below 0.5% of dry mass, and harvests that swing wildly with the season. Scaling up is rendered economically unviable.

Perhaps the most compelling incentive for this Botanical name theft lies not in local economical or industrial interest. Perhaps the Australian interest group's effort to claim the Acacia name lays in the brand recognition.

ECONOMIC SIGNIFICANCE COMPARED

Having shown that the original African thorn trees are emperors of the SKU economy (anchoring ingredient decks across pharmaceuticals, functional foods, cosmetics, and industrial chemistry), it's worth asking why the International Botanical Congress (IBC), and Australia's lobbying machine, ignored the elephant in the trademark room.

Search the world's intellectual property archives (trademark offices, patent banks, regulatory registries) and the pattern is unmistakable. Acacia gum appears in tens of thousands of packaged food, beverage, and supplement launch records from 2010–2024 in European and North America markets, according to Innova Market Insights and Mintel GNPD [8]. From Nestlé's soluble coffee to Mondelez's gummy candies, to GlaxoSmithKline's oral rinses, and Johnson & Johnson's pediatric syrups, the ingredient listing is crystal clear. "Acacia" isn't a garden shrub. It's an emulsifier. A prebiotic. A sign of trust and efficacy. It means one thing, and one thing only. It's the African tree with a global economic fingerprint.

So, when taxonomists reassign that name to a lineup of ornamental wattles and minor-league ingredients, they're not just reorganizing species. They're disrupting the language of industry. Rebranding the backbone of global commerce with something ornamental, and regional.

The Patents speak volumes

Sometimes, the real story isn't written in headlines or even in books. It's buried in the fine print footnotes of patents, the chemical codes of FDA filings, and the obscure corners of trademark registries. When it comes to Acacia gum, those quiet paper trails tell a thunderous truth: the world has already made its choice.

Search for "acacia gum" or its many synonyms ("gum acacia" OR "acacia gum" OR "gum arabic" OR "arabic gum" OR "E414" OR "E-414") and

ECONOMIC SIGNIFICANCE COMPARED

you will find an avalanche. Google Patents alone returns more than 100,000 search results in the USA and Europe alone for patents starting in 2010 [9]. These filings span encapsulation systems, tablet coatings, beverage emulsifiers, prebiotic blends, and biomedical gels. Ingredient after ingredient, product after product, all anchored to a gum exuded by African and Asian thorn trees.

The implication is striking. Behind every one of those documents is a formulator, a chemist, a legal team, and a product manager all taking for granted Acacia is the African/Asian thorn tree. In the secret language of commerce and regulation, Acacia already means something very specific. It's a trusted input with global recognition and centuries of safety.

Now try the same Google patent search for Australian species post 2010: ("wattle gum" OR "wattle extract" OR "wattle tannin" OR "wattle seed" OR "wattle oil" OR "wattle fiber" OR "wattle fibre"). Then switch wattle with any specific industrially significant Australian species like: mearnsii, melanoxylon, pycnantha, acuminata, dealbata, decurrens, mangium, auriculiformis, and salinga [10]. And what do you get? 388 search results in total. This means a ratio of Acacia gum to Australian species extract of at least 258:1 in the past 15 years only. This ratio alone should shame the IBC decision.

While African/Asian Acacias are embedded in formulations from cough syrups to nutritional drinks; Australian acacias remain within niche cosmetic novelties, artisanal flours, and boutique botanicals. This isn't merely a difference in product category. It's a difference in market identity and that's where the story gets even more consequential.

Regulatory filings aren't passive archives; They're active economic passports. When Nestlé registers a beverage blend that uses Acacia gum, or when GSK files a new oral care patent with Acacia gum, they're not just making products. They're building regulatory precedents and cementing Acacia gum as a safe, effective, and irreplaceable global commodity. These filings shape Codex standards, influence FDA GRAS notices, and guide WHO formulations. In all the above, the species they point to, almost without exception, is the African/Asian thorn tree. This brings us to the larger truth which is that brand dominance doesn't come from marketing alone. It comes from ubiquity, trust, and institutional memory. Acacia gum (real Acacia gum) is no longer just a natural substance. It's a documented global standard in thousands of filings and across dozens of industries. It is coded into the commercial DNA of the food, pharmaceutical, and nutraceutical sectors.

So, when Australia's lobby argues that their species deserve the name, the paperwork begs to differ. In the eyes of global regulation, Acacia already means something, and it doesn't have anything to do with a wattle.

SKU Disruption: If the IBC Got Its Way

Imagine you're the head of regulatory compliance for Nestlé or Coca-Cola. You've spent decades ensuring that every ingredient in your product line (every drop, flake, gum, or fiber) meets international standards. You live by codes: E-numbers, GRAS status, pharmacopeia references. You don't gamble with words, especially not with one as widely entrenched and globally trusted as "Acacia gum".

Since the International Botanical Congress (IBC) issued its controversial 2005 decision to reassign the Acacia name to Australian species, something telling has happened. Or rather, it hasn't. Neither the U.S. FDA (latest amendment 2023), nor the European EFSA (2019), nor Japan's Ministry of Health (MHLW, updated 2025), nor India's FSSAI (2023) have revised the name "Acacia gum". Not a single regulatory authority has followed IBC's lead. That silence speaks volumes about the name's embedded trust, built over decades of safety, utility, and trade integrity.

Even more surprising is the Codex Alimentarius, the international food code jointly developed by the FAO and WHO. It's the backbone of food law in over 180 countries, harmonizing everything from additive limits to ingredient names. Codex doesn't just regulate food; it sets the rules of the game for global commerce. Codex's current framework still defines "Gum Arabic (Acacia gum) (INS 414)" as the global benchmark for natural hydrocolloids, with a clear regulatory definition that supports its use as a bulking agent, carrier, emulsifier, glazing agent, stabilizer, and thickener. It is legally classified for use in confectionery, sauces, dressings, supplements, and beyond [11].

This isn't bureaucratic inertia. It's recognition of a fundamental truth that when food and pharma industries say "Acacia," they mean something very specific. They mean a gum from African thorn trees and not ornamental wattles from Australia. That meaning is encoded not only in law, but in safety records, lab trials, and generations of product reliability.

Now imagine this. The IBC decision is universally enforced everywhere from Codex to the FDA to Japan. Imagine the situation where "Acacia" now points to Australian species. The consequences wouldn't be academic. They'd be catastrophic. Let's walk through the wreckage. Suddenly, "Acacia gum" on a product label could legally refer to an entirely different class of ingredients. A new ingredient without GRAS status, toxicology profiles, or usage history. Food manufacturers would scramble to revise ingredients, label claims, and marketing language. Contracts would collapse. International supply chains would fracture and that's just the beginning.

Coca-Cola, Mondelez, Nestlé, PepsiCo, Ferrero, and Danone would be forced to relabel the unknown amounts of SKUs in their beverage and confectionery portfolios. Pharmaceutical companies would have to resubmit regulatory filings for every lozenge, tablet, or suspension containing Acacia gum. GS1 UK estimates the cost of a single SKU label change at £1,300–£4,000 ($1,600–$5,000) [12]. Multiply that by hundreds of thousands, and the numbers turn obscene. This doesn't even account for the damage to IP as hundreds of thousands of patents citing Acacia gum (and its synonyms) as a functional agent would suddenly be misaligned with regulatory definitions. Retailers and distributors would face recalls. Developing world manufacturers (those without regulatory war chests) would be destroyed. For the African and Asian farmers who have harvested and exported true Acacia gum for generations, the rebranding would erase hard won market access overnight. Their "Acacia" would vanish, renamed in the fog of a taxonomic power play.

Australia's defense, of course, is that a name change would cause them confusion. Hobbyists, students, and chemical industries using wattle wood, tannins or seeds might be baffled by a shift to "Racosperma. The honest reality is, however, that the confusion of a few hundred botanists, enthusiasts and industrial purchasing agents pales beside the economic devastation of relabeling the world's most trusted plant gum. The IBC scenario is a case study in regulatory overreach. It proposes to rewrite one of the most commercially trusted botanical names on Earth, without any regard for who pays the price.

When talking Commerce, Count the SKUs

The next time an international botanical congress debates the importance of a genus, perhaps it should leave behind the sterile allure of species counts and open something more revealing like the balance sheet of global commerce. Buried in barcodes, regulatory filings, and supply chain contracts lies a truth too large to ignore. The African/Asian Acacias are not just species, they are systems. They feed, they heal, they employ, and they connect the world.

It takes a certain kind of audacity to recast a narrow industry of tanning extracts and ornamental wattles as the rightful heir to the name Acacia, while quietly relegating the global supply chain of Acacia gum to a footnote. That is exactly what the IBC's decision enabled. There is, however, a test that is even more compelling than taxonomy. There is not a country on Earth where "Acacia gum" does not anchor at least a thousand SKUs. There is not a continent that wouldn't feel its absence in the form of product recalls, ingredient shortages, or disrupted formulations.

Now contrast that with the Australian wattles. They were certainly locally revered and ecologically significant in the past. However, they are not present in the lives of billions of global consumers. To frame Australia's claim as an economic necessity is to confuse national pride with global utility. Commerce has already spoken. Consumers have already voted. Walk the aisle of any supermarket in Paris, Nairobi, Tokyo, or São Paulo, and the answer is clear. The Acacia that matters is the one that nourishes, heals, stabilizes, and sells.

Irreplaceability, The Definitive Proof

In 1997, the United States drew a line in the sand. Comprehensive sanctions were imposed against Sudan, halting virtually all trade and financial transactions with the country. Interestingly, amid the sweeping restrictions, one curious exemption remained: Acacia gum arabic (INS 414). It was not an afterthought. It was policy explicitly carved out under the "food and medicine" exemption, codified at 31 CFR § 538.526 [13]. Why make such a distinction for tree sap?

The reasoning is that this was no ordinary ingredient. Acacia gum wasn't a boutique luxury; it was essential to the world's largest economy. The

Acacia gum exemption lasted two decades until sanctions were completely lifted on the country and relations restored. Through wars, policy shifts, and diplomatic freezes, the U.S. quietly allowed this one product to flow, uninterrupted. Not out of charity, but necessity. Beverage companies warned of product destabilization. Pharmaceutical firms cautioned against supply breakdowns and U.S. policymakers, balancing national security with commercial stability, listened. According to various studies, Acacia gum is widely valued for its unique functional properties, such as emulsification, film-forming, stability and extremely high prebiotic content. While alternatives like guar gum and xanthan gum are used, the review notes that formulating equivalent performance often requires blends or higher usage levels, potentially increasing complexity and cost [14].

Now place that against Australian Acacia offerings of Wattle seeds, tannins, timber, and wax. They don't hold global industries together. They don't dictate regulatory exemptions and are economically interchangeable. There's no crisis if they vanish and they certainly don't appear in the fine print of international sanctions law. That's the critical factor.

Acacia gum from the thorn tree isn't just functional, it's geostrategic. It's a rare plant product whose absence triggers panic in boardrooms and policy chambers alike. And when the world's largest consumer of Acacia gum (while implementing the world's strictest sanctions regime) still makes room for it, the message is clear. This African tree sap is uniquely irreplaceable. In the world of global commodities, that's the final word. Irreplaceability is not about origin; it's about impact and only one Acacia has earned that distinction.

The Real Difference: Australian Acacia vs. African/Asian Acacia

Sometimes, the simplest truths are obscured by the most technical arguments. Let's step down from the technical clouds of botanical taxonomy and ask a more human question. Who is actually using these products and what happens when their names are changed?

Start with the African/Asian Acacias and their signature Acacia Gum. The direct consumers are ordinary people, unaware of the subtle difference between Senegalia senegal gum and Acacia seyal gum, much less the hundreds of Australian Acacia species. To the average consumer, "Acacia gum" is not a scientific category but a promise. It's a promise to a mother buying prebiotic fiber for her child's digestion. A promise to a food manufacturer ensuring emulsification and shelf stability. A promise to a pharmaceutical formulator depending on a well-tested excipient for lozenges and rehydration powders. In these cases, labeling isn't a technicality. It's a matter of health, law, and public trust.

Now contrast that with the Australian wattle products of tannins, timber, and restoration mulch. These are handled by specialist industry professionals, sourcing raw materials based on chemical specs, not supermarket labels. If their bulk shipment of bark is labeled "wattle" today and "Raco" tomorrow, it changes nothing in the chain of safety or public exposure. At worst, a few invoices get reprinted. Nobody ends up in the hospital.

This is exactly why the IBC's decision to assign the name Acacia to the Australian lineage isn't just a semantic shift, it's a systems failure. It granted an International taxonomic clarity to a niche industry, and in the process introduced regulatory chaos into the global food and pharma industry where clarity is most critical. By favoring the interests of Australian forestry lobbies, the IBC inadvertently disrupted a global system built on consistency and safety. Suddenly, a name still recognized across

Codex listings, FDA approvals, and EU regulations became vulnerable to misinterpretation and mislabeling. The real-world consequence of this decision lies in increased procurement risks. In economic terms, the mislabeling of Australian wattles was a branding issue yet the mislabeling of African/Asian Acacias is a crisis. The IBC's decision didn't solve a scientific problem. It solved a political one. In doing so, it opened the door for the possible confusion of millions of direct consumers and the possible threatening of global public health.

References:

1. SPINS, LLC. Ingredient Census – Acacia Fibre/Gum Arabic, Natural Enhanced + MULO Channels, CY-2024. Chicago, 2025; Innova Market Insights. Digestive Health Ingredient Tracker: Acacia Fibre/Gum Arabic (E 414) in EU-27 and GCC Launches, 2019–2024. Report no. 24-062, April 2025; Nutrition Business Journal. "Prebiotic Fiber SKUs: Acacia Market Review," 2022.
2. Food and Agriculture Organization of the United Nations. Non-Wood Forest Products 11: Tropical Tannin Plantations. Rome: FAO, 2002, 7–11.
3. Grand View Research. Tannin Market Size, Share & Trends Report, 2023–2030. San Francisco, 2023; Data Bridge Market Research. Global Tannin Market – Industry Trends and Forecast to 2029. Chicago, 2022.
4. International Tropical Timber Organization. Market Information Service Quarterly Review 29, no. 2 (2024): 14–18.
5. "Black Wattle." TimberSearch ANZ. Forest & Wood Products Australia, 2024; Hoadley, R. "Black Wattle." Wood-Database.com, updated 2023.
6. United States Patent Application US20080311063A1. "Cosmetic Lash-Curling Composition." Published December 18, 2008.
7. Correia, Ana I., Inês Ribeiro, Joana Amaral, Maria A. Duarte, and Fernanda Carvalho. "Phenolic Composition and Antioxidant Properties of Acacia Species with Potential Industrial Applications." Plants 12, no. 19 (2023): 3450.
8. Innova Market Insights. Ingredient Insider: Acacia Gum Launch Tracker, 2010–2024. Arnhem: Innova, 2025; Mintel Group Ltd. Global New Products Database (GNPD), search query: "acacia gum," "gum acacia," "arabic gum," "gum arabic," "E414," "E-414."
9. Google Patents. Advanced search
10. Google Patents. Advanced search

11. Codex Alimentarius Commission. General Standard for Food Additives (GSFA), 47th Session. FAO/WHO, 2024.
12. GS1 UK. The Cost of a Label Change: A Guide for FMCG Brands and Retailers. London, 2020, 6–9.
13. Office of Foreign Assets Control. "Sudan and Darfur Sanctions." U.S. Department of the Treasury.
14. "Gum Arabic and Its Substitutes for Food Applications: An Appraisal." ResearchGate, 2024.

CHAPTER 13
ECOLOGICAL SIGNIFICANCE COMPARED

Australian Acacia species (especially wattles like Acacia mangium and Acacia mearnsii) were widely planted both in Australia (their native range) and globally for forestry, pulpwood, tannin, and land restoration. On the other hand, the African/Asian Acacias are used for products like Acacia gum, animal fodder, fuelwood, and reforestation. This chapter compares the environmental impact and sustainability of industries based on Australian wattles versus those using African/Asian Acacias.

We will assess the key sustainability metrics of both species. These metrics are soil health, water use, industrial waste, greenhouse gas (GHG) emissions, biodiversity, and invasiveness. We then explore the availability of eco-friendlier alternatives for similar end uses and conclude with a structured comparison of the overall ecological impacts. The goal is to determine whether the global net ecological impact of Australian wattles is positive or negative, and how it contrasts relative to its African/Asian counterparts.

Reforestation, Soil health and Desert Greening:

Picture two crews of tree planters standing at opposite edges of the world's degraded lands, each hoisting seedlings they swear are the silver bullet for soil fatigue and creeping desert. The first crew handles the Australian wattle (eg: Acacia mangium). These species were considered the quick solution for reforestation specialists. Drop them into a scorched grassland or a mine spoil and they rapidly perform a biochemical rescue act. They quickly fix atmospheric nitrogen with litter layers that jump start carbon accretion and soon the scorched soil begins to pulse with microbial life [1].

In their native continent, these wattles knit seamlessly into bush ecosystems, shoring up slopes and riding out bush fires without complaint. However, plant them on an existing biodiversity hotspot and they morph into ecological vandals. Their nutrient-rich leaf fall combusts habitats evolved for austerity. They will raise the pH, spike mineral nitrogen, and give weeds the energy needed to expand and conquer land previously belonging to native plants [2]. In a new foreign plantation, under a planted wattle canopy, islands of lush fertility form. It's alluring in the beginning but overtime they leapfrog across the landscape, eroding the very diversity that land managers hoped to heal.

Now cross to the opposite planting line, where Sahelian farmers cradle Acacia seyal and Acacia tortilis. These trees arrive with no missionary zeal; they are simply coming home. Their names headline the African Union's Great Green Wall species short-list because they stabilize dunes, furnish fodder, and shower the topsoil with organic matter while remaining obedient to the governance of native insects, browsers, and fire [3]. In Senegal or Burkina Faso, a farmer will dig small catchments into a crusted field, shield the A. seyal seedlings from animals, and soon watch herbs and millet reclaim ground that was scorched only months earlier [4].

On the Arabian flank, foresters plant A. tortilis not just for shade but for a hydraulic service few pumps can match. This Acacia's umbrella crown lifts scarce rainfall from thirty-meter-long roots, reduces wind scour, and through judicious pruning feeds animals without killing the mother tree. Hydrologists have measured what folklore always claimed which is fields sprinkled with these parkland Acacias double the rate at which stormwater sinks into Sahel soils and halve the sheet runoff that would otherwise race away [5]. There is even an expert thesis that the enormous belts of Acacia might even tweak local rainfall, though more proof is needed. What is certain is that micro-humidity climbs under every canopy, an oasis effect that science can already map [6].

Carbon economics further splits the two crews. The Australian wattles boast amazing carbon uptakes compared to African/Asian species. A tempting statistic for pulp companies and carbon traders. Yet those gains are rented, not owned. In 6-8 year rotations, much of that carbon is liquefied into black liquor or bleached into paper that decomposes, sending CO_2 back to the sky like a bounced cheque. Worse, when wattles set up shop in peat-rich Southeast Asia, they drain the water table, oxidize millennial carbon, and turn the carbon ledger back to negative. The Sahelian Acacias, by comparison, sequester a modest one to three tons of carbon per hectare per year, but they do it on land that previously stored almost nothing, and they never require the collateral of intact rainforest or peat. Coppiced every few years for fuelwood, they regrow from the stump. Lifecycle accountants count the released CO_2 as repaid by the regrowth dividend.

Even their modes of failure diverge. When Australian wattles overperform abroad, they do so because no local herbivore has yet learned the taste of their seedlings. More dangerously, the wattle's seedbanks linger for decades, and the very traits that once greened a moonscape keep on greening until diversity flattens. By contrast, when native African/Asian Acacia thickets thicken too far, Sahelian herders simply cut branches for fodder, fire sweeps in, or animals prune the exuberance. Therefore, for the African/Asian species, "Bush encroachment" is a feedback problem, not an alien invasion.

Restoration scientists therefore frame the decision less as a contest of growth rates and more as a choice between provisional nurses and permanent citizens. Plant a wattle and you get an expressway to shade and nitrogen, with border control costs that escalate over time. Plant the African/Asian Acacia and you must wait longer, but the tree slots into an ecological credit union already recognized by microbes, pollinators, and grazers.

Water Use and Hydrology

Imagine two thirsty landscapes and two rival clans of trees riding in to quench them. The first clan of Australian wattles (Acacia Mangium and its cousins) drive roots deep, pump water hard, and keep their evergreen canopies shimmering even in the dead of the dry season. The Australian species planted in Southeast Asia and Amazonian plantations confirm this reality. A single hectare of fast-growing wattles can siphon more soil water than the adjacent native forest and visibly flatten the seasonal rise and fall of the local water table [7].

This is, however, not the full story. Australian wattles in South Africa tell the follow up story of how roughly one third of all stream flow reduction credited to invasive plants nationwide is laid at the feet of these Australian wattles, the single largest taxonomic culprit [8]. These Australian species never shed their leaves in sympathy with the climate and since they hoard water year-round; they reduce base-flow even when rivers shrink to strings of puddles. This results in the slow erasure of groundwater away from aquifers that humans and plants share [9]. The result is a hydrological "hot spot", an oasis for the tree itself but a sinkhole for everything downstream.

This reality prompted South Africa's "Working for Water brigades" to chainsaw wattles first in their catchment restoration triage [10]. Farther north, the peatlands drained to make way for A. mangium plantations release millennial carbon and quickly exterminate local biodiversity. Farther west, in Peru and Brazil, these plantations result in shifting water flow peaks forward in the calendar so that water arrives when nobody needs it and vanishes when everything is thirsty [11].

The second clan (the African/Asian thorn trees: seyal, tortilis, nilotica) present an ecological happy land. Their brilliance lies in timing. In Burkina Faso, infiltration experiments show that soil beneath a mature A. seyal absorbs the season's first monsoon downpour twice as fast as bare ground. The tree's deep roots fracture the hardpan that once acted like a ceramic roof, while its fallen leaves keep soil pores open and ready to take in water [6]. After the rains, these savanna Acacias shift to austerity. Their leaves yellow and drop, the trees enter drought dormancy, and they coast through the dry "hunger season" on minimal water, never drawing more than the arid landscape can spare. [12].

In Symbiosis with these African/Asian Acacias, farmers dig small, half-moon shaped hollows around the trunks of these thorn trees. When it rains, the hollows catch the stormwater, letting it soak in slowly instead of rushing away and eroding the soil. Satellite pictures back this up showing fields dotted with Acacia seyal/tortilis holding moisture deeper in the ground, so millet can keep reaching water long after plants in bare fields have withered. Some scientists even wonder whether long belts of these trees might draw in a bit more rain by sending tiny, moisture-rich particles into the air, a "biotic pump". That idea is still being tested, but local measurements already show clear benefits. The shade under the trees is cooler, the air is more humid, and afternoon breezes carry a faint, pleasant gum scent.

The difference is almost ironic. Australian wattles march in as empire builders. They grow fast and grab as much water as they can; even using the water that will be needed later. In new habitats, this leaves nearby plants short of moisture. If wattles spread too far, they can lock a whole watershed into a permanently drier state that lasts for decades, even after the trees are cut down. African/Asian Acacias act more like community

ECOLOGICAL SIGNIFICANCE COMPARED

bankers. They take only small amounts of groundwater, drop leaves that add nutrients back into the soil, and slow their growth before a drought begins. When these thorn trees become too dense, a problem called "bush encroachment", people can usually fix it with simple measures such as grazing animals, setting controlled fires, or cutting extra trees. These quick actions restore the local water balance without the need for large, costly eradication programs.

None of this implies that Australian wattles lack utility. On nutrient poor mine spoils, their deep, thirsty roots act like pumps, dragging hidden minerals up to the surface and jump-starting new plant life for the next wave of species. The problem starts when those same water hoarding trees are planted in regions already running short on rainfall. Introducing wattles there is like writing a check your watershed can't cash; they pull tomorrow's water to build today's fast greenery. Studies show that a dense stand of wattles can cut local streamflow in half [8], and the landscape may stay drier for years even after the trees are removed.

Farmers in the Sahel take a different approach. They line their contour bunds (low earth ridges that slow runoff) with Acacia tortilis or Acacia seyal. One neat row of these local acacias can double the amount of rainwater that soaks into the soil [6]. When the long dry season arrives, their crowns drop leaves and go dormant, conserving moisture instead of losing it to the air [12].

The numbers make the contrast hard to ignore. One tree line of Sahelian Acacias boosts infiltration, while a thicket of wattles drains streams. That is why South African water regulations blacklist wattles as invasive, while Sahelian farmers count their native Acacias among their most valuable assets.

In summary, for the wattles; plant too many, too dense, too far from home, and the long-term repercussions tips toward drought, igniting a feedback loop of deeper roots and drier streams. For the African/Asian thorn trees, the ecological impact is slower, gentler, and more sustainable.

That is why restoration agencies now embed hydrological tests when selecting species, knowing that every species carries an invisible water meter at its root tip. In the arithmetic of desert greening, the aim is not merely to add green but to preserve the blue in the river. With this metric, the evidence suggests that native African/Asian seyal and their kin are a net positive while expatriate Australian wattles all too often run in the negative.

Industrial waste, pollution

When comparing the Australian Wattles and African/Asian Acacias in this category; imagine two industrial assembly lines that depart into entirely different waste geographies. On the Australian side, the belt starts in regimented rows of Acacia mangium or black wattle. Logs rumble into the mega mills of Sumatra, Kalimantan, and Espírito Santo, where they are shredded, cooked in white liquor, and sluiced out again as kraft pulp. Every stage leaves a fingerprint. Logs are cooked in white liquor to produce kraft pulp, and if washing is incomplete, residual black liquor carries into bleaching. Chen et al. (2017) found that this residue supplies more than 34% of the effluent's COD and increases AOX levels, highlighting a clear fingerprint of contamination in wastewater [13]. Picture the mountains of wood bark residue (thousands of tons per site) stacked like damp matchsticks because the high-bark fraction of Australian logs is worthless for fine pulp and only some of it can be of any industrial use beyond that point [14].

Downstream, factories that boil the bark of Acacia mearnsii to make "tannin extracts" capture only about one-third of its valuable polyphenols.

The other 70 percent ends up as soggy, spent fiber. Plant managers have three options for that waste: bury it in landfills, burn it in open pits, or press it into experimental briquettes and biochar that are still looking for a reliable market [15]. The wastewater is another headache. When those pipes carry more liquid than the undersized treatment ponds can handle, the overflow slips into nearby creeks. Drainage canals built for plantations discharge sulfuric and acidic water into creeks. Villagers report foul smells and declining fish populations, an environmental hazard repeatedly documented in the peatland regions of Sumatra and Borneo [16].

Shift the frame to the African/Asian thorn belt and the background noise changes from turbines to hand tools. Here Acacia gum is tapped from African/Asian Acacia trees that dot Sahelian parklands. Farmers score the bark with a curved knife at the start of the dry season; weeks later they flick off marble size resin lumps. The resin lumps are then sun-dried and carried to the village market. The tree is never truly damaged, so canopy cover persists, roots keep stitching soil to subsoil, and field surveys in Kordofan and Darfur show that households earning gum income nurse extra seedlings therefore boosting tree density beyond that of untapped rangeland.

These African/Asian woodlands act like an ecological happy land. The deep roots lift moisture, leaf litter drips organic matter and mineral nitrogen, and umbrella crowns blunt sandstorms. This cycle results in slowing wind erosion and stalling the desert front [17]. It's no wonder the World-Bank "green-growth" projects now pay Mauritanian and Nigerien villagers to extend these groves as part of the Great Green Wall [18].

The Gum's processing is very clean with minimal waste. The collected Acacia resin lumps undergo hand sorting, bark flaking, and sieving. No sulfurous digesters, no bleach towers, and every reject flake is biodegradable. Lifecycle audits show that forest harvested Acacia gum carries a greenhouse gas tab less than half of chemically modified starch while offering superior emulsifying properties in many applications [19].

The contrast grows starker when we follow the fate of the tree's wood. Across the communities of East Africa, African/Asian Acacias provide the everyday fuel of tea kettles and evening stews. Households gather fallen branches and selectively cut stems for charcoal in a traditional woodland management technique called coppicing with the additional benefit of resetting the tree's vitality clock every three to five years [20].

This sustainable wood production technique means that each combustion cycle from chopped off wood is matched by a regrowth pulse. The result is a carbon ledger that hovers near neutral, unlike diesel or grid power drawn from distant coal [21]. Studies from Kenya's Baringo District show wood-fuel stocks holding steady for decades, soil cover intact, local biodiversity humming, and no long-haul trucking of timber required [22]. Even small-scale leather tanning from the boiling of Acacia nilotica pods generates little residue as the spent mash becomes compost or feed, not a hazardous slurry [23].

This comparison is especially revealing. Australian industry, scaled for global pulp and leather demand, tumbles into a waste-to-resource ratio so lopsided that entire national programs (South Africa's Working for Water and Brazil's invasive-species task forces) are mobilized to mop up bark mounds and detox effluent ponds. The African/Asian thorn economy, by contrast, is circular. Resin harvesting that cumulatively produces thousands of Acacia gum tons ends up positively influencing the overall ecology. In addition to the Acacia Gum industry, sustainable wood harvesting results in tree regrowth while the small-scale leather tanning byproducts cycle into soil fertility.

Both Australian and African/Asian Acacias produce hundreds of thousands of product tonnage, but only one does so while holding its ecological credit card close to a zero balance. The moral is clear; industries built on Australian wattles must spend heavily on closed-loop effluent plants, biochar pilots, and landfill fees merely to keep pace with the waste they generate. Industries built on African/Asian seyal, tortilis, and nilotica invest that same capital in farmer cooperatives and regeneration; converting each bead of gum or stick of fuel into a dual dividend of rural income and restored savanna. The products may have shared a genus on the label, but the downstream footprints could not be further apart.

Fodder and fire related risks

The story of Australian wattles and wildfires plays out on two different stages. In the first stage, picture endless rows of Acacia mangium planted on the peat flats of Kalimantan. These trees drop oily leaves year-round, pull groundwater with deep roots, and fill the forest floor with evergreen litter. In a normal dry season, that is already risky. Add an El Niño drought, drain the peat for drainage canals, and the ground itself turns into tinder. When a spark lands, the tidy rows act like wicks where flames race up the crowns, then burrow into the carbon-rich peat below. The fire can smolder for months, pumping gray haze across Southeast Asia and grounding flights half a continent away [24].

Stage two, move west to South Africa or Portugal's sandy coast and the pattern repeats. Plantations of Acacia mearnsii or A. dealbata shed long ribbons of bark and heaps of dry leaves. Lightning or a tossed cigarette is all it takes. Once lit, the wattles burn hot, the heat from the fire triggers the buried seeds of these trees (seeds that can survive decades) and this results in carpeting the ash with thousands of new seedlings [25]. Each fire they feed becomes a built-in recruitment drive for the next, even hotter blaze.

Fire behavior models back up what crews on the ground already know.

Replace a patchwork of low, native shrubs with a wall of wattle and the fire intensity can double. Native seed banks cook in the heat, biodiversity drops, and the land resets for yet more wattles, locking in a cycle of bigger, faster fires [26]. In short, where these Australian wattles move in, wildfire risk doesn't just rise; it accelerates. These flames do not respect fences. They leap into adjacent pine, into subsistence gardens, and finally into the insurers' ledgers. Transferring the ecological mess into economic losses.

Look north to the Sahel and you get a very different picture. Instead of dense, flammable walls of trees, you see scattered thorn trees standing like watchtowers across farm fields. Their timing is perfect. Just as the summer monsoon arrives and crops of millet and sorghum need full sun, the thorn trees shed every leaf. Sunlight pours onto the seedlings. When the harvest is over and the land turns to stubble in the long dry season, the same trees leaf out again, shading the soil, slowing evaporation, and dropping protein-rich pods for livestock. The benefits run deeper, literally. Taproots bore into the subsoil, pull up hidden moisture and minerals that result in dropped leaves, pods, and eventually manure (after animals browse for fodder). The cycle renovates topsoil with richer organic matter, improved structure, and higher porosity [27].

Biodiversity tags along for the ride. Midday shade cools the micro-climate, buried dung pats sprout clover for bees, thorny branches cradle bird nests, and animals can strip leaves without killing the trees. The net effect is a park-like landscape that boosts, rather than crowds out life in every direction.

Fire is still a fact of life in the Sahel, yet the local thorn trees help control it instead of fanning the flames. During the rainy season, when lightning strikes are most common, these Acacias stand bare, so there is little fresh "green" fuel to ignite. The trees are scattered, not packed together, which breaks up continuous bands of dry grass. Year-round grazing and browsing by livestock and wildlife trim off extra branches, keeping fuel loads low. If a grass fire does sweep through, it rarely climbs into the treetops, and the Acacias' thick bark shrugs off the heat.

Compare that picture to the landscapes planted with Australian wattles and you get the exact opposite effect. Brought in for pulpwood and tannin, wattles act like built in fire starters. They drain wetlands, pile up dry litter, and then quickly reseed after a blaze. More fire means more wattles, a loop that keeps the land drier and riskier each year. Cleaning up the mess is expensive as governments have to spend millions on chainsaws, herbicide-spraying helicopters, and peatland repairs.

Sahelian Acacias flip that script. Farmers treat each tree as a living firebreak and a quietly working pump that lifts nutrients from deep soil layers. Every mature crown provides shade, livestock fodder, and steady drops of organic matter that builds richer topsoil. One system pushes hidden costs onto taxpayers and emergency crews. The other stores value in the field itself, turning every tree into a little insurance policy against crop failure, dust storms, and runaway fires.

Understanding these tradeoffs before the first seed is sown may be the difference between landscapes that dehydrate people and feed flames, and landscapes that quench people while taming flames.

Greenhouse Gases and Carbon Balance

Australian wattles often look like climate champions at first glance. In Vietnam, Acacia mangium plantations have stacked up 110–170 tons of carbon per hectare in just sixteen years, and the steady rain of leaves keeps adding more to the soil [28]. On tired grasslands, places that used to hold less than eight tons of carbon, this same tree has turned the scrub into eighty-ton carbon stores within two decades [29]. On paper, it reads like a miracle fix.

Look closer and the shine fades. Most of these plantations run on six- to eight-year pulp cycles. When the logs head to the mill, their carbon ends up in disposable paper or burned as boiler fuel, so the bulk of that "captured" carbon returns to the atmosphere almost as fast as it was stored. Only a sliver ends up in long-lasting products [30]. The picture turns even darker when plantations push onto peat-swamp forests. Draining the peat lets oxygen in, and centuries-old carbon starts to oxidize. Flux towers in Sumatra show the land losing up to eighty tons of CO_2-equivalent per hectare each year [31]. When you factor in the drainage canals leaking dissolved organic carbon into the sea, each harvested hectare quickly becomes a net emitter.

Fire completes the storm. Once the peat dries, it burns easily, adding thick smoke to the air and keeping Indonesia high on the list of the world's land-use greenhouse gas sources [32]. Even clearing ordinary rainforests for an Australian wattle planation exacts a heavy price. Those ordinary rainforest sites once locked 250–400 tons of carbon per hectare. Young Wattle stands can't pay back that debt for decades, often not within the plantation's business timetable at all [33].

So, the same tree can be a hero or a hazard. Plant wattles on carbon-poor wasteland and they build a valuable stockpile (assuming you can control their spread). Plant them over ancient carbon (peat swamps or primary forest) and they turn into accidental arsonists, setting off chains of emissions that dwarf their short-term gains.

Move the camera from Southeast Asia to the wide, dusty Sahel and the story flips. Local thorn trees (species like Acacia seyal and A. tortilis) aren't being grown for pulp mills and they are not planted on ancient rainforests. They survive leaf by leaf, pod by pod, even while goats nip at their branches and sandstorms scrape their bark. Each year they add just 1 - 3 tons of carbon per hectare on land that held almost zero carbon to begin with, every ton is pure gain. Since there was never a dense rainforest bulldozed to plant them, there's no hidden "carbon debt".

Farmers in Niger and Chad simply protect wild seedlings or plant fresh ones on crusted plains that agronomists had written off. Over time those flats turn into "living carbon vaults". Branches pruned for firewood grow back quickly. The carbon released when that wood burns is "prepaid" by new growth the tree will make in the next season. Families swap free fuel for kerosene or imported charcoal, saving cash and cutting fossil emissions. Meanwhile, the standing trees give livestock shade and fodder, offer flowers to pollinators, cool the soil, and slow down hot winds. These bonuses never show up on a carbon spreadsheet but will make a big difference when the next drought rolls in.

Biodiversity, Invasiveness and Ecosystem Impacts

Australian wattles were exported on the promise of fast shade and faster timber, yet their ecological resume abroad reads like a list of unintended consequences. Research shows that monoculture wattle plantations support significantly fewer wildlife species than primary or secondary forests. One study in Brazil's Atlantic Forest found that "wattle invaded second growth stands" stored more carbon but had just one-third to one-sixth the tree diversity (taxonomic, phylogenetic, and functional) of non-invaded forests [34].

Roraima, Brazil, offers the next cautionary chapter. A. mangium, planted for pulp, now sprouts thickly in nearby savanna and flooded forest, edging out native herbs and seedlings faster than managers can pull them [35]. This is no outlier. At least 23 Australian wattles have naturalized in more than fifty countries causing dense, nitrogen-fixing thickets that dim the light, rewrite soil chemistry, and tilt fire regimes toward higher heat [36]. The Cape fynbos, famous for endemics, is half-strangled by the Australian A. saligna and A. cyclops; their walls of evergreen foliage halve local plant richness and lay down tinder that pushes every wildfire into the red zone [37].

In Portugal and Spain, the Australian wattles ride sea breezes inland, their nutrient-rich litter elevates soil pH and mineral nitrogen just enough to favor yet more wattles and suppress the dune scrub adapted to sand and scarcity [38]. Seed rain from plantation perimeters compounds the problem. Millions of hard-coated seeds fall each year, viable for decades, guaranteeing reinvasion after every mechanical clearing or herbicide round [39]. Ecosystem services crash in tandem. South African studies tie wattle invasions to shrinking streamflow, poorer forage, and plummeting pollinator counts, costs that echo well beyond the initial studies [40].

Why are Wattles Invasion Over-Performers?

Australian wattles arrive with a kit of evolutionary tools including turbocharged growth, fire-primed seeds, nitrogen fixation, and drought tolerance that give them a permanent head start once freed from their native herbivores and pathogens. Their seeds can sit in the soil for fifty years [41] and then germinate en masse after the first hot burn, creating self-reinforcing armies that bulldoze botanical diversity. South Africa's black-wattle saga illustrates the scale. Australian Acacias now occupy more than a million hectares, choking rivers and climbing mountain catchments critical to municipal water supplies [42]. A. saligna and A. longifolia, introduced for dune control, have laced the Cape Floristic Region with fuel and fertilizer, shifting an entire fire-adapted biome onto a hotter, faster burn cycle [43].

Scientists warn that today's "stable" A. mangium plantations in Asia and

ECOLOGICAL SIGNIFICANCE COMPARED

the Americas merely hide an invasion debt. Seed banks accumulate quietly, ready to erupt decades after the last ribbon-cutting ceremony. The economic drag is as relentless as the ecological one. South Africa's Working for Water program now spends billions of rand on saw crews, herbicides, and seed-feeding weevils, yet must revisit the same hillsides repeatedly because biological agents only dent, never erase, the seed supply.

Contrast that with the native thorn trees across Africa and western Asia. African/Asian Acacias have anchored savanna and scrub ecosystems for millennia. Their open crowns and seasonal leaf drops furnish a living scaffold for grasses, herbs, bees, birds, and grazers.

Global invasion databases list more than twenty Australian wattles among the woody invaders on earth while a few of them are in the top 100 list worst invader list. On the other hand, none of the of the African/Asian Acacias make the top 100 list [44]. The rare counterexample proves the rule when African/Asian nilotica was planted in northern Australia as a shade tree. The absence of browsing giraffes and goats allowed it to explode across six million hectares of Queensland rangeland, forcing Australia to declare its former guest a Weed of National Significance and pour millions into mechanical, chemical, and biological control [45].

The same cautionary tale echoes in the Horn of Africa, where Prosopis juliflora (another Mimosoid outsider) has morphed into a green plague, reminding land managers that any nonnative dry-zone tree planted without safeguards can flip from asset to adversary [46]. Even so, such cases are dwarfed by the global tally of Australian wattles. Within their own sub-humid belt, African/Asian Acacias are more likely to suffer from "bush encroachment". The management fix is rotational grazing and controlled burns, not national eradication campaigns [47].

Industries built on African/Asian Acacia's gum tapping, dry-season fodder, and parkland agroforestry therefore reinforce indigenous biodiversity. Farmers who protect gum trees find more pollinators in their millet fields. Herders lop branches for feed and see faster grass rebound. Birds nest in thorny crowns while wild ungulates browse alongside cattle. All of this occurs without triggering the domino effects that were followed by a wattle invasion. The same cannot be said for Australian wattle enterprises, which too often exchange short-term pulp or tannin gains for long-term ecological liabilities. As discussed, these include diminished species richness, altered nutrient loops, hotter fires, and perpetual control costs. The lesson is as old as ecology itself. A tree's value is written not just in its growth rings but in the history of the ground it grows upon and the web it weaves with every neighbor, friend or foe.

Ecologically Friendlier Alternatives

Pulpwood and Timber: Australian wattle estates locked into the pulp supply chain are discovering that they are no longer indispensable. After wilt disease decimated Acacia mangium blocks in Sumatra and Kalimantan, the companies APP and APRIL simply pivoted. Slotting clonal Eucalyptus pellita and E. urophylla grandis into the same rotations. Field trials showed near-equal pulp yields and far less pathogen pressure. While the eucalypts' short life span seeds cut long-term invasion risk in half.

Researchers have recently expanded their methods for producing fiber. Instead of planting only one type of tree, they now mix faster growing native trees. Recent research in Sarawak shows that mixed plantations of Neolamarckia cadamba and Gmelina arborea perform well in early growth and pulp quality assessments sufficient for small-scale paper operations [48].

Beyond trees, researchers are exploring other plant materials. Bamboo, harvested every three years, is now a major source of material for paper and rayon production in China and India. Life-cycle analyses in China show that industrial bamboo products like flooring and panels can be carbon-negative when sustainably harvested and responsibly processed [49]. Other assessments show that substituting just 10–20% of wood pulp with non-wood fibers like bagasse, wheat straw, or reed can significantly reduce greenhouse-gas emissions and land-use impacts compared to conventional wood pulp [50].

By contrast, African/Asian Acacias were never drafted into high-volume pulp works. Their timber is secondary to gum, fodder, and fuelwood. All of which are harvested in extensive parklands where regrowth matches offtake and seed banks are a feature, not a bug. This is the case because local communities tap resin or prune branches rather than clear entire trunks. There is little pressure to find "replacements". The trees already operate within an agro-ecological loop, enriching soils and stocking carbon without generating the waste streams or invasion debts that are a feature of Australian wattle plantations.

The bottom line: Wattles can be swapped for eucalypts, native hardwood mixes, bamboo, or recycled fiber with minimal technical pain. African/Asian thorn trees remain irreplaceable precisely because their greatest value is ecological, not industrial.

Tannin Production: Black wattle tannin sits at the center of an uneasy equation. The bark of Acacia mearnsii delivers a reliable, condensed extract for heavy leather. Nonetheless, each ton comes with ecological consequences. Invasive A. mearnsii increasingly chokes South Africa's waterways and has overtaken cork-oak hillsides in Portugal and Spain [51]. Tannery chemists concede that botanical equivalents already exist. Quebracho heartwood from Schinopsis species matches the wattle's astringency [52].

Newer attention pivots to lower risk feedstocks. Sweet Chestnut coppices in Europe supply high-purity hydrolysable tannin while doubling as multipurpose timber lots [53]. Valonia oak acorn cups keep a centuries-old, small-scale trade alive in the eastern Mediterranean [54], and Andean farmers now ship increasing volumes of Tara pod extract for chrome-free formulations, all without spawning weedy thickets.

Inside South Africa, managers have flipped the liability into a resource by stripping bark from existing wattle infestations for tannin and burning the wood for bioenergy, a combined "extractive invasion control" that removes seed trees even as it feeds local kilns.

Across the Sahel and Indian subcontinent, the story runs quieter. Rural tanners boil pods and bark from African/Asian Acacia in village vats, then compost the spent fiber. The scale is cottage, the waste biodegradable, and the trees, integral to dry-land parklands, regrow every pruning cycle, posing no invasion threat at home or abroad.

Where black wattle demands remote plantations, extraction mills, and perpetual containment budgets, African/Asian species fold tannin into existing agro-ecologies, enriching soils and feeding livestock between harvests. The comparison suggests a simple calculus. To meet leather's future without multiplying ecological debt, the industry can blend chestnut, Tara, valonia, and locally managed extracts while turning current wattle thickets into a dwindling, not expanding, raw-material bank.

Acacia Gum: Australian wattle gums have lately been pitched as a home-grown alternative to African sourced Acacia gum, yet laboratory trials show that even the best Australian wattle exudates still fall short of the archetypal African/Asian Seyal and Senegalia resin in all its various applications. Other equivalent alternatives to Acacia Gum result in food formulators blending in extra surfactants and sacrificing label simplicity.

For example, Guar gum (E 412) thickens admirably but boosts viscosity so sharply that beverage makers must add separate emulsifiers to keep flavor oils dispersed. Xanthan gum (E 415) is derived from lab fermentation processes and delivers a limitless pseudoplastic material yet likewise needs a dose of gum Acacia fractions or modified starch to match the flavor/oil stability that comes effortlessly to Acacia gum. Locust bean gum, carrageenan, and agar shine in gelling dairy textures but falter when asked to carry essential oils at parts-per-million levels required for soft drinks.

Sustainability metrics sharpen the contrast. Guar draws praise for its arid adaptability, yet its annual cycle still demands tillage, fertilizer, and mechanized harvests that lift its embodied energy score [55]. Xanthan's footprint rises or falls with the carbon intensity of its fermentation input. Not only is the carbon source in the fermentative process responsible for one-third of the production costs, different carbon sources and stress conditions could affect the molecular structure. Meaning that more green innovation for fermentation will produce xanthan gum with different properties and therefore different emulsification capabilities [56].

By comparison, Acacia gum tapped from wild African/Asian Acacias posts minimal greenhouse gas emissions, less than half those of guar on an equal-function basis. This is because harvesting Acacia gum relies on nothing more energy intensive than a curved knife and the patience to tap the bark and later sun dry collected lump resins. All while the Acacia woodland quietly stores carbon and grips fragile Sahel soils. The socio-economic dividend is equally lopsided. Gum revenues are often the only cash stream in remote drylands and therefore motivate communities to guard and reseed their parklands rather than clear them, an effect documented from Kordofan to Zinder.

Analysts agree no single hydrocolloid matches the full tool kit of African/Asian Acacia gum. Replacement schemes require bespoke blends that raise cost and viscosity penalties. The pragmatic path is not to swap out this ingredient but to fortify its supply chain, support farmer cooperatives, plant new thorn belts, and certify provenance. All of this will ensure that the already irreplaceable resin in formulation remains sustainable in origin.

Global Net Ecological Impact of Australian Acacias: Positive or Negative?

Finally, evaluating all the sustainability metrics together, we ask: Is the global net ecological impact of Australian Acacia species positive or negative? Below is a structured comparison, contrasting Australian wattles (grown worldwide) with African/Asian Acacia Species, and weighing their pros and cons:

- **Soil Health:** Australian wattles: Improve soil fertility and aid reforestation on degraded sites (positive), but in pristine ecosystems they enrich soil beyond natural levels, facilitating invasive undergrowth and altering microbiota (negative).
  African/Asian Acacias: Improve and restore soils in degraded drylands (positive), essentially with no soil-related downsides when planted in native range (they maintain natural soil balance).

- **Water Resources:** Australian wattles: Often deplete water. Their high transpiration and interception reduce river flow and groundwater, especially when invasive (A. mearnsii et al. cutting catchment yield by ~30+%). Their use on peatlands necessitates draining, causing long-term hydrological damage (negative).
  African/Asian Acacias: Generally, water-neutral or positive as their deep roots enhance infiltration, and stands can stabilize climate

moisture (e.g. Acacias in Sahel help retain rainwater and may induce more rain locally) They are adapted to water scarcity, so they don't overly tax the environment. By preventing desertification, they protect water cycles (positive).

- **Industrial Pollution:** Australian wattles: Large-scale monocultures feed industries that produce significant waste (pulp mill effluents, bark residues). Mismanagement can lead to water and soil pollution (negative).
  African/Asian Acacias: Gum tapping and collection along with other local-scale usage (fodder, fuelwood) produces negligible pollution as these processes are natural or low-tech with no toxic by-products (positive).

- **GHG Emissions & Carbon:** Australian wattles: When used to afforest truly barren lands, they can sequester carbon and benefit the climate (positive); however, the dominant reality has been clearing carbon-rich forests/peat and replacing them with short-rotation wattles, causing a net surge in emissions. Invasive fires in wattle thickets also release carbon. Thus, from a global perspective they are negative.
  African/Asian Acacias: Planting these trees is part of climate-change adaptation in the Sahel as they sequester carbon (albeit modest amounts) and avoid emissions by anchoring soils (positive). They are not linked to any large-scale carbon emissions spike. They help communities mitigate and adapt (e.g. providing renewable wood fuel reduces the need to burn fossil fuels or cut down distant forests).

- **Biodiversity:** Australian wattles: Outside Australia, net negative for biodiversity as they simplify habitats, outcompete native species, and sometimes form monocultures, reducing overall species richness. Only in their native Australian context are they neutral or beneficial as part of the native flora.

African/Asian Acacias: Net positive for biodiversity in their ecosystems. More Acacias mean more food/habitat for native fauna, and greener landscapes in degraded zones (as shown by wildlife returning to restored Acacia gum areas). They increase landscape heterogeneity which generally boosts biodiversity.

- **Invasiveness:** Australian wattles: High invasive potential and a clear global negative. They have established invasive populations on every continent except Antarctica, with severe ecological and economic costs. Managing these invasions will require ongoing resources for decades.
African/Asian Acacias: Low invasive risk when kept in indigenous regions (neutral/positive). They coevolve with the ecosystem and do not become pests in those contexts. Caution is only needed when moving them outside their native range (Not common).

- **Ecological friendlier alternatives:** Australian wattles: once central to pulp, tannin, and gum industries, are increasingly replaceable. Eucalypts, bamboo, and non-wood fibers now outperform them in yield and sustainability, while tannin demand is shifting to chestnut, tara, and valonia oak.
African/Asian Acacia gum: remains unmatched in emulsification, sustainability, and socio-economic impact. Unlike wattles, African/Asian Acacias thrive within agro-ecological systems, regrowing naturally, enriching soil, and supporting livelihoods without invasive fallout. Where wattles need control budgets and industrial infrastructure, African/Asian Acacias deliver ecological and economic value through low-impact harvesting and community-based stewardship. Their replacement is neither possible nor wise.

Considering all these factors, the global net ecological impact of Australian wattle species has been more negative than positive. While they have provided economic benefits (fast wood, tannin, and restoration of some degraded lands), those come at a steep environmental cost when wattles are planted for industrial or reforesting plantation purposes. Their global toll is one of invasions, water resource loss, native biodiversity decline, and significant greenhouse gas emissions in most industrial cases. In their native environments (Australian landscapes), wattles are valuable members of the ecosystem, but those native benefits do not translate globally.

Therefore, the "ecological importance" of Australian wattles according to proposal 1584 is a mere illusion as the entire globe is spending billions to rid itself of them. Their introduction outside the Australian continent is an ecological catastrophe. The IBS's decision to grant the name to this globally recognized ecological pest is a true affront to the name and legacy of the True Acacia.

References

1. L. S. Koutika, H. J. Rainey, and I. Tchichelle, "Changes in Soil Organic Carbon and Total Nitrogen under Acacia mangium Plantations in Congo," Annals of Forest Science 71, no. 5 (2014): 495–504.
2. Stephanie G. Yelenik, William D. Stock, and David M. Richardson, "Ecosystem-Level Impacts of Invasive Australian Acacias in the Fynbos Vegetation of South Africa," Annals of Botany 101, no. 5 (2008): 187–192.
3. African Union Commission, Great Green Wall Implementation Status and Way Ahead to 2030 (Addis Ababa: AU, 2020).
4. Jules Bayala et al., "Interactions between Trees and Crops in Agroforestry Systems in the Sahel," Agriculture, Ecosystems & Environment 297 (2020): 106882.
5. Jules Bayala et al., "Water Infiltration under Two Acacia Species in Semi-Arid West Africa," Soil Use and Management 30 (2014): 48–56.
6. Paolo D'Odorico et al., "Progress in Research on the Sahel Greening Phenomenon," Nature Sustainability 4 (2021): 649–657.
7. David L. McJannet et al., "Evaporation and Tree Water Use in a Tropical Rain-Fed Plantation and a Temperate Native Forest," Hydrological Processes 17, no. 20 (2003): 407–422.
8. David C. Le Maitre et al., "Impacts of Invasive Australian Acacias on Water Resources in South Africa," South African Journal of Science 96, no. 9/10 (2000): 651–656.
9. Peter J. Dye and Cheryl Jarmain, "Water Use by Black Wattle (Acacia mearnsii): Implications for Stream-Flow Reductions," Water SA 30, no. 4 (2004): 43–52.
10. Brian W. van Wilgen et al., "The Working for Water Programme in South Africa: A 20-Year Review of Its Implementation for Controlling Invasive Alien Plants," Bothalia 46, no. 2 (2016): a2112.

11. Silvio F. B. Ferraz et al., "Effects of Plantation Forests on Streamflow in the Amazon and Atlantic Forests," Forest Ecology and Management 310 (2013): 347–356.
12. Hailu Kassa et al., "Farmer-Managed Natural Regeneration for Restoring Degraded Drylands in Ethiopia: A Systematic Review," Land 10, no. 7 (2021): article 724.
13. Tiantian Chen et al., "Influence of Residual Black Liquor in Pulp on Waste-Water Pollution after Bleaching Process," BioResources 12, no. 1 (2017): 2031–2039.
14. Nursinta A. Rosdiana et al., "Characterization of Bark Extractives of Different Industrial Indonesian Wood Species for Potential Valorization," Industrial Crops and Products 108 (2017): 121–127.
15. Juliana Schultz et al., "Tannin-Industry Waste-Derived Porous Carbon: An Effective Adsorbent from Black Wattle Bark for Organic Pollutant Removal," Sustainability 16, no. 2 (2024): article 601.
16. "Deforestation and Fires Persist in Indonesia's Pulpwood and Biomass Plantations," Mongabay, May 23, 2025. https://news.mongabay.com/2025/05/deforestation-and-fires-persist-in-indonesias-pulpwood-and-biomass-plantations/.
17. Charles Fagg and I. A. Stewart, "Restoring the Gum Arabic Belt in Sudan with Local Communities," Arid Lands Research and Management 34 (2020): 155–173.
18. World Bank, "Gum Arabic Supports Green Growth in Mauritania," July 6, 2015.
19. "Reinforcement Strategies and Applications in Food Preservation: Gum Arabic Review," Trends in Food Science & Technology 132 (2023): 194–208.
20. Joseph K. Githiomi and Bernard Chikamai, "Woodfuel Situation, Outlook and Implications for Sustainable Forest Management in Kenya," Discovery and Innovation 24, Special Issue (2012): 32–43.

21. Emmanuel N. Chidumayo, "Woodfuel and Charcoal Production in Sub-Saharan Africa: Issues and Concerns," Energy for Sustainable Development 17, no. 4 (2013): 312–320.
22. R. Mendum and H. Glen, "Sustainable Woodfuel Harvesting in Baringo," World Agroforestry Centre Field Report (2009).
23. B. H. Dzowela et al., "Tannins of Acacia nilotica: Local Extraction and Applications in Small-Scale Leather Curing," Tropical Science 42 (2002): 40–47.
24. Mongabay News, "Perpetual Haze: Pulp Production, Peatlands, and the Future of Fire Risk in Indonesia," 2018.
25. Leonel J. R. Nunes et al., "The Impact of Rural Fires on the Development of Invasive Species: Analysis of a Case Study with Acacia dealbata in Casal do Rei," Environments 8, no. 5 (2021): 44.
26. "Fire Regimes and Management Options in Mixed Grassland–Fynbos Landscapes," Fire Ecology 20 (2024): article 61.
27. Jean-Marc Boffa, Agroforestry Parklands in Sub-Saharan Africa, FAO Conservation Guide 34 (1999).
28. Nguyễn Hải Phạm et al., "Biomass and Carbon Storage in an Age Sequence of Acacia mangium Plantation Forests in Southeastern Vietnam," Forest Systems 29, no. 2 (2020): e009.
29. Rodel D. Lasco et al., "Biomass and Carbon Distribution on Imperata cylindrica Grasslands and the Potential of Reforestation," Agroforestry Systems 59 (2003): 23–32.
30. Gary D. Paoli and Lisa M. Curran, "Carbon Accounting and Pulpwood Rotations in Southeast Asia," Forest Ecology and Management 365 (2016): 101–112.
31. Aljosja Hooijer et al., "Carbon Dioxide Emissions from an Acacia Plantation on Peatland in Sumatra, Indonesia," Biogeosciences 9 (2012): 617–630.
32. "The Carbon Brief Profile: Indonesia," Carbon Brief, March 19, 2019; Adhityani Arga, "Indonesia World's No. 3 Greenhouse Gas

Emitter—Report," Reuters, June 4, 2007.
33. Rodel D. Lasco and Florencia Pulhin, "Carbon Debt from Converting Tropical Rainforests to Tree Plantations in Southeast Asia," Global Change Biology 17 (2011): 226–237.
34. Fabio A. R. Matos et al., "Invasive Alien Acacias Rapidly Stock Carbon, but Threaten Biodiversity Recovery in Young Second-Growth Forests," Philosophical Transactions of the Royal Society B 378 (2023): 20210072.
35. Joseph W. Veldman and Francis E. Putz, "Long-Distance Dispersal of Invasive Acacia mangium into Savanna and Forest in Roraima, Brazil," Biotropica 43, no. 6 (2011): 758–761.
36. Agnaldo Aguiar Jr. and Reinaldo I. Barbosa, "Invasion of Acacia mangium in Amazonian Savannas Following Planting for Forestry," Plant Ecology & Diversity 7, no. 4 (2014): 359–369.
37. David C. Le Maitre et al., "Impacts of Invasive Australian Acacias on Biodiversity in South African Fynbos," Journal of Ecology 99 (2011): 100–109.
38. Elizabete Marchante et al., "Invasive Acacia longifolia Changes the Soil Properties of Portuguese Coastal Dune Ecosystems," Plant and Soil 320 (2009): 237–248.
39. Steve Csurhes and Greg Neilson, Invasive Potential of Acacia mangium in Plantation Landscapes (Queensland: Department of Agriculture and Fisheries, 2018).
40. David C. Le Maitre et al., "Water and Ecosystem Services in Areas Invaded by Alien Trees," South African Journal of Science 103 (2007): 467–472; B. W. van Wilgen, Tsungai A. Zengeya, and David M. Richardson, "A Review of the Impacts of Invasive Alien Species in South Africa," Biological Invasions, Stellenbosch University Repository, 2021.
41. Marianne Strydom et al., "Invasive Australian Acacia Seed Banks: Size and Relationship with Stem Diameter in the Presence of Gall-Forming Biological Control Agents," PLOS ONE 12, no. 8 (2017):

e0181763.
42. David C. Le Maitre et al., "Impacts of Invasive Australian Acacias on Biodiversity, Water and Fire Regimes in South Africa," South African Journal of Science 107, nos. 5–6 (2011): 1–11.
43. Stephanie G. Yelenik, William D. Stock, and David M. Richardson, "Ecosystem-Level Impacts of Invasive Acacia saligna in South African Fynbos," Restoration Ecology 12, no. 1 (2004): 44–51.
44. David M. Richardson and Marcel Rejmánek, "Trees and Shrubs as Invasive Alien Species—A Global Review," Diversity and Distributions 17, no. 5 (2011): 788–809.
45. Queensland Department of Agriculture and Fisheries, Prickly Acacia (Vachellia nilotica) Control Manual (State of Queensland, 2021).
46. Ross T. Shackleton et al., "The Invasion Ecology and Management of Prosopis juliflora in Africa: A Synthesis," Ambio 46 (2017): 407–421.
47. David Ward, "Decades of Vegetation Change in the African Savanna: Bush Encroachment as an Ecological Indicator," Ecological Indicators 99 (2019): 19–27.
48. Foong Kee Chai et al., "Growth and Pulp Properties of Mixed Neolamarckia cadamba and Gmelina arborea Plantations in Sarawak," Industrial Crops and Products 169 (2021): 113666.
49. Piet van der Lugt et al., "Environmental Impact of Industrial Bamboo Products," Journal of Cleaner Production 231 (2019): 1176–1184.
50. Ning Li et al., "Life-Cycle Assessment of Wood versus Non-Wood Pulp Supply in China," Journal of Cleaner Production 313 (2021): 127781.
51. Paulo C. Zonetti, Renata C. de Souza, and Jussara M. Campos, "Acacia mearnsii De Wild.: Uses, Chemical Composition, and Management of an Invasive Species," Revista Brasileira de

Biociências 15, no. 2 (2017): 95–106.
52. Alejandra Rodríguez-Vargas et al., "Characterization of Tannin Extracts from Selected Native Quebracho (Schinopsis balansae) for Leather Applications," Industrial Crops and Products 120 (2018): 195–203.
53. R. Zanuttini and F. Scacchi, "Chemical Characterisation and Industrial Uses of Sweet-Chestnut Tannin Extract," Journal of the Society of Leather Technologists and Chemists 103 (2019): 1–7.
54. Yoshikazu Yazaki, "Utilization of Oak (Quercus spp.) Valonia for Tannin Extraction," Holzforschung 61 (2007): 173–178.
55. Raj Kumar et al., "Life Cycle Assessment of Guar Gum Production in India," Journal of Cleaner Production 265 (2020): 121787.
56. Renata A. Trindade, Adriel P. Munhoz, and Carlos A. V. Burkert, "Impact of a Carbon Source and Stress Conditions on Some Properties of Xanthan Gum Produced by Xanthomonas campestris pv. mangiferaeindicae," Biocatalysis and Agricultural Biotechnology 15 (2018): 167–172; Meirielly Jesus et al., "Corncob as Carbon Source in the Production of Xanthan Gum in Different Strains Xanthomonas sp.," Sustainability 15, no. 3 (2023): 2287.

CHAPTER 14

FUTURE ECONOMIC AND ECOLOGICAL TRENDS

Global Acacia gum sales originate almost exclusively from African/Asian species belonging to the Acacia Seyal and Senegalia. According to Grand View Research, the market was valued at approximately $526 million USD in 2024 and is expected to grow at a compound annual growth rate (CAGR) of 6.4%, reaching nearly $880 million USD by 2032 [1].

This established market now underpins a larger transformation across the global consumer packaged goods landscape. The current tens of thousands of SKU's relying on Acacia Gum will surely explode as its adoption becomes mainstream. Analysts expect three converging forces to significantly increase this number over the next decade.

First is the prebiotic shift. By 2035, most global beverage brands are expected to launch "functional hydration" or "symbiotic" lines. Acacia Gum, as a soluble, clean-label dietary fiber and emulsifier, stands to benefit directly. Even capturing a conservative 8% of the projected $20.5 billion USD prebiotic ingredient market by 2033 could treble the number of beverage and supplement SKUs using Acacia gum [2].

Second is the ongoing clean-label normalization. Food manufacturers are removing carrageenan, modified starch, and synthetic emulsifiers in favor of multifunctional natural ingredients. Acacia gum (designated E 414) offers emulsification, film formation, and fiber enrichment in a single compound. As it replaces multiple ingredients across frozen desserts, RTD coffees, plant-based beverages, and savory sauces, it may contribute additional tens of thousands of SKUs globally.

Third, the pharmaceutical and nutraceutical sectors are increasingly using Acacia gum nanoparticles in delivery systems. Given its GRAS and European Pharmacopoeia (Ph. Eur.) status, Acacia Gum bypasses the regulatory delays affecting newer polymers. A 5% penetration into the $2.5 billion USD plant-based excipient market could substantially increase its presence across prescription and OTC formats including many delivery technologies from probiotic capsules to peptide sachets. [3].

Specialized applications such as edible films for 3D printed foods, solvent-free drug carriers, biodegradable capsule shells, and thermal-stable vegan food systems (e.g., printed meats, ice cream foams) will likely drive additional diversification, especially through social media trends. Overall, we project another addition of 50,000+ Acacia gum dependent SKUs by 2035 in the EU and North America alone. This represents a significant increase in SKUs.

Moreover, this projected SKU explosion is more than a numerical milestone, it represents a market moat. Each line on a retail buyer's spreadsheet translates to shelf space, regulatory filings, and consumer-facing identity. In a marketplace where name recognition and label continuity are critical to trust and compliance, this proliferation of SKUs locks the term "Acacia gum" into global systems. In fact, it's already locked in. Reformulating or renaming it would entail enormous financial and regulatory cost, disrupting not only commercial contracts but public safety protocols tied to ingredient traceability.

In contrast, Australian Acacia species (such as Acacia mangium and Acacia mearnsii) serve largely bulk commodity sectors. In 2024, Australia exported approximately 4.093 million bone-dry metric tons of hardwood chips at a value of $262 AUD per bdmt, generating an estimated $720 million USD [4]. Additionally, Acacia mearnsii bark is a principal source of vegetable tannins. The global market for tannins is projected to reach $4.9 billion USD by 2032, with black-wattle extracts comprising roughly one-third, or about $1 billion USD [5]. However, increasing environmental scrutiny, emission liabilities from peatland plantations, and widespread fiber substitution trends are expected to reduce the combined Australian Acacia revenue forecast significantly. The future forecast for Australian Acacias in a world moving toward ecofriendly solutions is bleak at best.

The multifunctionality of Acacia gum further strengthens its market resilience. It is an amphiphilic, cold-water-soluble polymer that simultaneously provides emulsification, film-forming capacity, with a high soluble-fiber content (approximately 90%) all in a single FDA approved ingredient. No known alternative replicates this combination, whereas eucalyptus, bamboo, bagasse, quebracho, and synthetic alternatives can readily substitute Australian wattle-based fibers and tannins.

Even more important in this discussion is the fact that none of the top regulatory bodies use the new botanical names forced on the "True Acacia" trees. The FDA, EFSA, and Codex all insist on using the term Acacia for Seyal extracts and therefore this means all new patents and SKUs will continue bearing the name in continuity with the past 2,300 years.

Environmentally, the case is even clearer. African Acacia species are integral to the African Union's Great Green Wall initiative, which aims to restore 100 million hectares of degraded land and sequester 250 Mt CO_2 by 2030 [6]. Field studies confirm associated benefits such as temperature reductions and yield improvements of 30–40%. Conversely, Australian Acacia plantations in Indonesia emit roughly 20 Mt CO_2-eq annually due to their location on drained peat soils [7]. In South Africa, invasive species including Acacia mearnsii lower regional water availability by an estimated 8%, causing annual economic losses of $875 million USD [8]. The European Commission has formally designated Acacia saligna as an Invasive Alien Species of Union Concern, legally requiring member states to contain or eradicate it [9].

Policy measures mirror these realities. While Indonesia enforces a moratorium on new peatland concessions and South Africa funds active eradication through the "Working for Water" program, African Acacia cultivation benefits from carbon-credit mechanisms that offer payments of approximately $10 USD per ton of CO_2 sequestered; creating a regenerative economic incentive.

Finally, consumer protection and label integrity reinforce the case for retaining the name "Acacia gum" for the African/Asian Species derived products. This nomenclature is widely understood by food regulators, retailers, and consumers across product categories. This entrenched label recognition spans infant formulas, nutraceuticals, medical syrups, nutritional supplements and pharma formulations. Renaming the product could invalidate centuries of safety documentation, patents and regulatory approvals, jeopardizing supply chain continuity and exposing markets to adulteration risks.

In conclusion, the African/Asian species have emerged not only as economic drivers but as future cornerstones of sustainable agriculture, nutrition, and pharmaceutical innovation. By contrast, Australian wattles face shrinking market space, rising ecological liabilities, and declining relevance. The commercial, ecological, and regulatory evidence align. The right to carry the Acacia name into the future belongs to the species whose products are indispensable, high-value, and increasingly foundational to global health.

References:

1. Grand View Research, Gum Arabic Market Size, Share & Trends Analysis Report, 2024 (San Francisco: Grand View Research, 2024).
2. Straits Research, Prebiotic Ingredients Market Size, Share & Trends Report 2024–2033 (Pune: Straits Research, 2024).
3. Grand View Research, Pharmaceutical Excipients Market Size, Share & Trends Analysis Report (Plant-Based Segment), 2024–2030 (San Francisco: Grand View Research, 2023).
4. IndustryEdge, Woodchip Market Briefing—February 2024 (Melbourne: IndustryEdge, 2024).
5. Market.us, Tannin Market Size, Share, Growth Trends 2023–2032 (New York: Market.us, 2023).
6. Claudia Joubert, "Africa's Great Green Wall Gets New Life," Time, March 12, 2024.
7. David L. A. Gaveau et al., "Carbon Emissions from Peatlands Converted to Acacia Plantations in Indonesia," Nature Climate Change 6 (2016): 202–206.
8. Willem J. de Lange and Brian W. van Wilgen, "An Economic Assessment of the Contribution of Biological Control to the Management of Invasive Alien Plants and to the Protection of Ecosystem Services in South Africa," Biological Invasions 12, no. 12 (2010): 4113–4124.
9. European Commission, "Commission Implementing Regulation (EU) 2019/1262 of 25 July 2019 Updating the List of Invasive Alien Species," Official Journal of the European Union, July 26, 2019.

CHAPTER 15

HIDDEN IN PLAIN SIGHT – THE MOST OBVIOUS COUNTERARGUMENT

The Greek noun ἄκανθα (akantha) denotes a thorn or spine. Hellenistic botanists coined the diminutive ἀκακία (Akakía) for the Nile thorn-tree, embedding the root akanth- and thus retaining the semantic feature "sharp". Classical Greek botany embeds the concept of thorniness in the very terms that generated the modern generic name Acacia. The same root later sprouts in modern taxonomic tags such as Acantho-cephala ("spiny head") and Acantho-cereus ("spiny candle cactus") [1]. Classical writers used akantha generically but saved a special diminutive, Akakia (ἀκακία), for one particular thorn-tree of the Nile valley whose gum tightened wounded flesh and whose pods dyed leather a stubborn black. The name describes "Among the Egyptian thorn-trees is the Akakia, armed with many sharp points; its juice is powerfully binding" [2]. Pierre Chantraine points out how the syllable akan- wanders through Greek words for thorns, spines, and pricks [3]. Herodotus notes that Egyptians used planks from a thorny tree (διὸ ἀκάνθῃ for boat construction) [4] and Galen warns pharmacists to wash the gum lest fragments of spine remain [5].

Morphological surveys show that thorniness is the norm in the African/Asian clades historically called Acacia with 95 % of these species possessing paired stipular spines or prickles.[6] Thus, etymology (akantha → Akakia → Aqaqiya →Acacia), historical pharmacognosy (succus acaciae), and modern phylogenetic evidence converge on a single conclusion: the diagnostic feature encoded in the name (thorniness) is retained in the African/Asian lineages but largely absent from the now-typical Australian wattles. The taxonomic realignment therefore inverts the primary morphological signal embedded in two millennia of botanical and medical literature.

In contrast, the Australian wattle masquerading as Acacia s. str. after the 2005 International Botanical Congress re-typification comprises ~1,100 species, fewer than 2 % of which bear spines [7]. Only nine of which carry anything sharper than a pruning scar.

According to the Acacia Heist masterminds, the semantic ship has now sailed. A name that meant a thorn tree in Egypt for twenty-three centuries is now appropriated by thornless Australian wattles. It is as if the Olympic committee suddenly decided that "boxing" should cover only matches in which the contestants promise never to punch. The irony is botanical, but the logic is political. The Hellenistic ghosts must be smirking. Somewhere in a papyrus scented afterlife, Dioscorides is turning to his scribe: "Take note, Pedanius: the pointy plant we named for its points has, in bureaucratic eternity, lost the points but kept the name".

In short, Acantha begat Akakia, which Darwinian bureaucracy begat a genus almost perfectly lacking in Acanths. If anyone still doubts the mismatch, go hug an Australian wattle (no gloves required) and then try the same with a Nile thorn. Your skin will supply the final, irrefutable evidence needed.

It seems the dictionary now needs a disclaimer worthy of a Simpson episode:
Acacia (n.) [< L. acacia < Gk. akakia "thorny Egyptian tree"]
Terms and conditions apply: definition void in Australia, where the so-called "thorny" trees are thornless!

References:

1. Henry George Liddell and Robert Scott, A Greek–English Lexicon, 9th ed. (Oxford: Oxford University Press, 1940), s.v. "ἄκανθα."
2. Theophrastus, Enquiry into Plants 4.2.1–3.
3. Pierre Chantraine, Dictionnaire étymologique de la langue grecque (Paris: Klincksieck, 1968), 38, s.v. "akanth-."
4. Herodotus, Histories 2.96, trans. G. C. Macaulay (London: Macmillan, 1890).
5. Galen, De simplicium medicamentorum facultatibus 8.128, in Opera Omnia, vol. 12, ed. Kühn (Leipzig: 1826).
6. J. H. Ross, "The Thorny Acacias of Africa," Bothalia 10 (1971): 455–68.
7. D. J. Murphy, "Spinescence in Australian Acacia," Australian Systematic Botany 18 (2005): 133–44

CHAPTER 16

IF ONLY AUSTRALIANS ATE REAL ACACIA GUM

Welcome to Australia the land of sweeping plains, wallabies, weird venomous creatures, and some of the grumpiest guts in the developed world. We all know about the skin cancer rates, the sunburn, and the sharks. However, what the tourist brochures forget to mention is that millions of Australians are losing their battles daily; not on Bondi Beach, but in the intimate trench warfare of their own intestines.

Australia is in the midst of a full-blown gastrointestinal health crisis. The culprit is fiber. Or to be precise, the spectacular lack thereof. I'm not talking about mainstream fiber. I'm referring to the complex, microbiome-transforming, anti-inflammatory, immune-boosting prebiotic fiber. The kind found in the world's greatest, most proven, most tragically misunderstood gut healer: real Acacia gum. The gold-standard exudate from Acacia seyal and Senegalia.

The tragedy, dear reader, is that in a land that desperately needs a gut health revolution, the very word "Acacia" is being abused, misapplied, and commercialized into meaninglessness. Due to a mix of botanical bureaucracy, and regulatory lobbying; millions of Australians with irritable bowels, inflamed colons, and sad microbiomes can be sold the wrong stuff, under the right name.

Let's descend now, into the lower GI tract of the Acacia confusion. Have a seat, but maybe not for too long as constipation is rampant in these parts.

Down Under, In Trouble

The numbers are, well, hard to digest. A major multinational study published in Gastroenterology using the Rome IV diagnostic criteria found that 29.6% of Australians met the threshold for at least one FGID diagnosis (nearly one in three adults surveyed) [1]. That's about 9 million people waking up every morning to wonder whether their breakfast will start a civil war in their lower intestines. This aligns with clinical experience across Australia, where general practitioners and gastroenterologists increasingly report that functional gastrointestinal disorders (FGIDs) now form the majority of outpatient GI complaints. Unlike inflammatory bowel diseases (IBD), which show up clearly in pathology; FGIDs are rooted in the gut–brain axis and are frequently exacerbated by stress, diet, and microbiome imbalances.

The symptom burden is not only personal but economic. A multicenter audit in south Australia published in 2024 revealed that patients with FGIDs waited an average of 70 days for diagnosis and spent nearly six months in specialist care, often with little resolution [2]. This represents a significant drain on public health resources and lost productivity, especially since many patients are referred multiple times across specialties. Many of these individuals do not receive formal diagnoses, reflecting both the stigma and clinical uncertainty still surrounding conditions like IBS and functional bloating. What makes this more concerning is the frequent co-morbidity with mental health disorders. Research continues to highlight a bidirectional relationship between gut dysfunction and psychological stress, meaning patients often face a double burden of poorly understood digestive symptoms along with anxiety or depression [3].

Organic GI disorders such as IBD are also on the rise. A 2025 audit from Crohn's & Colitis Australia showed that while IBD affects fewer individuals than FGIDs, the lack of a national patient registry, uneven access to treatment, and long diagnostic delays mean Australia's IBD response is far from optimal [4]. Both functional and organic GI disorders require better public health investment, patient education, and dietary interventions.

Australia also records about 15,600 new cases of bowel (colorectal) cancer each year [5], making bowel cancer the fourth-most diagnosed malignancy overall after prostate, breast and lung cancer (and just behind melanoma). It is the second leading cause of cancer death, ranking only below lung cancer [6].

Here's a stat for health ministers to quote over their fake fiber enriched toast. A 2023 YouGov-commissioned survey for Bowel Cancer Australia found that a large proportion of Australians experience persistent bowel symptoms such as bloating, constipation, abdominal pain, or stool changes at least once a week [7]. Gastroenterologists and GPs alike are seeing an "epidemic of gut disorders", and in a twist so ironic it could make a Wattle tree weep, much of this misery is linked to diet.

If you want to know why Australians are in such a pickle, don't look to ancient genetic curses. The real culprit is a catastrophic lack of fiber diversity and poor prebiotic intake. National nutrition survey data show that the typical Australian adult manages only about 22 g of dietary fiber a day, short of the targets set by health authorities (25 g for women, 30 g for men). Most of that modest fiber ration comes from the usual suspects (wheat-based breads and breakfast cereals) while fruit, vegetables and legumes contribute far less than guidelines recommend [8].

Here is the interesting part; while Australians choke down psyllium husk and nibble on oat bran in the name of "fiber", they're missing out on the most effective, least celebrated, and most clinically studied prebiotic of all: real Acacia gum. Yes, the same African/Asian Senegalia and Acacia Seyal exudate that's been keeping American and European nutritionists busy incorporating it into their new diet resolutions.

Fiber? What is that?

Australians, like most Westerners, grew up thinking of "fiber" as a single, mysterious entity. Something vaguely brown, rough, and tasteless that doctors nag you to eat after the age of 40. The national conversation on gut health is dominated by images of "bran flakes", "wholemeal" and, if you're feeling brave, maybe a teaspoon of chia seeds in the morning.

However, the real magic of gut health isn't in bland grains. It's in the complex web of prebiotic fibers that nourish the diverse, hungry armies of bacteria in your colon. No fiber does this better, more reliably, or more gently than real Acacia gum (yes real and not wattle), also known as "Acacia fiber" in honest markets.

Hundreds of clinical trials, systematic reviews, and global expert consensus statements have now shown that Acacia gum (specifically from Senegalia and Acacia seyal) is an unrivaled prebiotic powerhouse [9]. It ferments slowly, promoting the growth of good bacteria (think Bifidobacteria and Lactobacillus), while minimizing gas and bloating which is a rare trick among fibers. Its excellence stems from its gradual fermentation in the colon resulting in the effective release of short-chain fatty acids (acetate, butyrate and propionate), rather than a single gas-laden surge. Elevated colonic butyrate is widely credited with supporting epithelial integrity, dampening low-grade inflammation, and providing a protective milieu that observational and mechanistic studies link to lower colorectal-cancer risk [10]. It is, in short, everything an Aussie colon is crying out for.

Contrast this with the typical "fiber" found in Australian pantries. Most are insoluble, poorly fermented, and do little to encourage microbiome diversity. The result? A bloated, undernourished population of bacteria that are, if anything, as bored as the consumers feeding them. Thanks to the Acacia Heisters, when an Australian with IBS or chronic constipation stares at the "Acacia" supplement aisle, they might find a bottle filled not with real Acacia gum, but with the resin of some local wattle. Offering little more than false hope and maybe even a medical complication.

The greatest joke in this entire affair is that the single most effective and research backed prebiotic in the world sits ready to save Australia's guts. The same population that needs it the most is now prime prey to false advertising.

The Foundation is set

Several Australian wattles, most notably Acacia georginae ("Georgina gidgee") and Acacia decurrens are flagged in national toxic plant registers and have been linked to livestock poisoning and occasional human adverse reactions [11]. This is where the "Acacia" branding scam becomes more than a labeling technicality; it becomes a public health risk. Australian consumers, desperate for gut relief, can buy "Acacia" supplements with the reasonable expectation that they're getting Acacia gum, the same fiber used in clinical trials, international medical guidelines, and recommended by international regulatory bodies. Instead, they may be ingesting extracts that have never passed a single gut health study, contain no prebiotic oligosaccharides, and might even trigger problems bigger than the ones they're trying to escape.

The Joint FAO/WHO Expert Committee on Food Additives (JECFA) has evaluated safety and set an "ADI not specified" for that specific African Acacia Gum and not for wattle exudates. Australian interest group's theft of the name Acacia has therefore placed their own constituents in the mercy of false advertising.

Not all Acacias are created equal. Some make fiber, some make tannin, and thanks to the Australian lobbying, they all make a great set of confusion.

Mislabel It and They Will Suffer

An Australian therapeutic Goods Administration (TGA) compliance review completed 76 compliance reviews of listed complementary medicines in 2020-21. They found that 40 of those reviews (53% of determinations) verified at least one breach of labelling, presentation, formulation or evidence requirements [12]. Given the huge enthusiasm among citizens of the developed world for Acacia gum and its prebiotic benefits. We foresee an avalanche of regulatory scandals driven by false advertising. The Australian botanical lobby has put the gut health of the developed world (including their own citizens) in jeopardy.

Real Acacia for Real Guts

Now, let's flip the script. Imagine a parallel universe, one in which the Australian food and supplement market was honest, scientifically literate, and immune to botanical ego. In this world, "Acacia" on a label would mean only one thing: genuine Acacia gum from Africa, with all its documented health benefits.

This isn't fantasy, it's the current practice in Europe, The Americas, Japan, India, China and the Middle East, where "Acacia" on an ingredient list is strictly regulated and must refer to the real Acacia. They have not even changed the naming of the ingredient from the original Acacia. The result? Consumers get a proven prebiotic, doctors can recommend products with confidence, and the national gut health profile improves accordingly. In fact, multiple clinical trials have shown that regular supplementation with Acacia gum:

- Increase beneficial gut bacteria,

- Decreases gut inflammation and symptoms of IBS,
- Promotes regularity and reduces constipation,
- Improves markers of immune health and metabolic well-being [13].

If just 10% of Australians with gut complaints switched to a daily 10-gram dose of real Acacia gum, the country could save millions in healthcare costs, reduce prescription drug use, and possibly even lower colon cancer rates. This will be a win for public health and national productivity.

For us to reach this goal, the food regulatory bodies should support us in our bid to reclaim the name in the IBC for the original thorn tree. Regardless of the Botanical nomenclature work that needs to be done.

References:

1. Ami D. Sperber et al., "Worldwide Prevalence and Burden of Functional Gastrointestinal Disorders: Results of Rome IV Global Epidemiology Study," Gastroenterology 160, no. 1 (2021): 99–114.e3.
2. Jane Mathias et al., "Functional Gastrointestinal Disorders in South Australia: A Multicentre Audit," Internal Medicine Journal (2024).
3. Neurogastroenterology & Motility Editorial Team, "Functional Gut Disorders and Their Psychiatric Comorbidity," Neurogastroenterology & Motility (2023).
4. Crohn's & Colitis Australia, State of the Nation in IBD Report (Melbourne: Crohn's & Colitis Australia, 2025), https://crohnsandcolitis.org.au.
5. Cancer Australia, Colorectal (Bowel) Cancer Statistics (Australian Government, 2023).
6. Australian Institute of Health and Welfare (AIHW), Cancer Data in Australia 2023, Cat. no. CAN 149 (Canberra: AIHW, 2023).
7. Bowel Cancer Australia, The State of the Nation's Gut Health: Bowel Cancer Australia & YouGov Survey Report 2023 (Bowel Cancer Australia, 2023).
8. Australian Bureau of Statistics, Australian Health Survey: Nutrition First Results—Foods and Nutrients, 2011–12.
9. W. Calame et al., "Gum Arabic Establishes Prebiotic Functionality in Human Intervention Study," British Journal of Nutrition (2008).
10. Henrike M. Hamer et al., "Review Article: The Role of Butyrate on Colonic Function," Alimentary Pharmacology & Therapeutics 27, no. 2 (2008).
11. Rosalind Dowling and John W. McKenzie, Plants Poisonous to Livestock in Queensland, revised ed. (Queensland: Queensland Department of Agriculture and Fisheries, 2016), 24–26 (entries for Acacia georginae and A. decurrens).
12. Therapeutic Goods Administration, Compliance Review of Listed Medicines: 2020–2021 Annual Report (Australian Government Department of Health, 2022).
13. Lamis Kaddam et al., "Gum Arabic Reduces C-Reactive Protein and Oxidative Stress in Patients with Sickle Cell Anemia: A Randomized Controlled Trial," Clinical Nutrition 36, no. 2 (2017).

CHAPTER 17

THE NEW ACACIA DECEPTION

When Australia's botanists secured the name Acacia for their continent's wattles at the 2005 Vienna Congress, they weren't merely settling a taxonomic quarrel. They were seizing a powerful marketing brand with deep consumer resonance. Generations of food, beverage, pharmaceutical and even cosmetics companies have built global product lines around the "Acacia" name, most prominently Acacia gum (E 414), which alone underpins tens of thousands of SKUs worldwide. By retaining Acacia for the 960-odd Australian species rather than reassigning the wattles to Racosperma, Australian producers could ride that existing brand equity for all manner of native wattle gums.

Acacia gum is the ingredient that keeps the word "Acacia" on the back panel of thousands of packaged foods. Market-tracking data from Netherlands Ministry of Foreign affairs and CBI analysis of Mintel Global show that between 2018 and 2022, more than 1,700 new European food and beverage products were launched containing Acacia gum. These product launches are especially concentrated in the confectionery (30%), beverages (24%), and bakery (20%) categories, reflecting the ingredient's core role in delivering texture, stability, and fiber content in everyday packaged goods. Major multinational brands including Nestlé, Mondelez, and Ferrero rely on Acacia gum in both flagship and functional food lines. Its plant-based, allergen-free, and clean-label-friendly profile makes it attractive to manufacturers responding to rising consumer demand for natural ingredients. Moreover, Acacia gum's classification as a soluble dietary fiber with prebiotic effects is fueling its inclusion in "gut health" and wellness products across the continent [1]. They are of course referring to the true Acacia gum, from the real Acacia thorn tree.

Unfortunately, by keeping the Acacia name with the Australian clade, local exporters could credibly label new hydrocolloid extracts as "Acacia gum", leveraging decades of downstream acceptance. This is dangerous and very confusing for millions of consumers internationally who depend on Acacia gum. The infrastructure for future false advertising has already been placed. How likely is this to happen? I will let you decide.

Dawn of the Fake Acacia

Research into novel Australia Acacia exudates illustrates the name's commercial potential beyond established gums. A 2024 Food Hydrocolloids paper on gidyea resin reported that polysaccharides exuded by Acacia cambagei produced continuous films and stabilized citrus-oil emulsions "comparable to or better than" Codex-grade Acacia gum, highlighting commercial promise for novel Acacia exudates beyond the traditional Sahel sources [2].

Australia's floral emblem (Acacia pycnantha-golden wattle) produces an exudate that researchers now tout as a potential "home-grown" hydrocolloid. Early ethnobotanical accounts record the sweet gum as a minor bushfood among southeastern Aboriginal groups, eaten fresh or dissolved in water. Building on that legacy, a 2023 pre-print argued that the exudate should qualify as a "traditional food" under Food Standards Australia New Zealand (FSANZ) rules, citing historical consumption and preliminary compositional data showing high arabinogalactan content [3]. Domestic labels describe the product as "for culinary experimentation" rather than a regulated food additive, side-stepping FSANZ approval but limiting uptake to niche markets. Under any other genus name these innovations would face steeper branding hurdles. However, as "Acacia" products, they tap into an existing confidence in E 414 and its kin.

Acacia decurrens ("green wattle") yields a thin, pale exudate that Indigenous Australians and early settlers occasionally used as a jelly-making agent. The gum can substitute for Acacia gum in small-batch fruit jellies, thanks to its basic gelling and stabilizing capacity. The main impediment so far is its low-yield, and brittleness.

Acacia victoriae (prickly wattle) has entered the margins of hydrocolloid research. Ethnobotanical records describe its pale, glassy gum being chewed fresh or dissolved in water by desert peoples of central Australia, alongside the better known use of its protein-rich seeds. A Rural Industries R&D Corporation scoping study sampled the gum in 2013 but concluded that average yields (under 40 kg ha^{-1} even in good seasons) were "too low and irregular for viable commercial milling" [4].

As you can see, university laboratories have officially supplied the technical narrative [2]. Press releases from the research teams highlight Australian gum's ability to suspend citrus oil in low-sugar beverages. All of this sets the foundation for craft soda and low-alcohol gin makers seeking clean-label emulsifiers. Pilot batches produced in Queensland have already appeared at specialty food expos under the trade name Australian "Acacia gum".

Historical precedent lends authenticity to the pitch. Colonial shipping manifests from Western Australia show small lots of "brown gum" exported to Britain during the 1880s. Records show that several colonies experimented with exporting wattle exudate as "Australian gum-arabic" but the venture faltered on both quality and cost. The record concludes that despite isolated shipments in the 1840s, "the high price of labor and the inferior quality of much of the gum render the industry unprofitable" [5].

The 2005 retypification thus weaponized the Acacia name for Australian interests. By keeping Acacia Australian, local industries could tout "pure Australian Acacia" gums, seeds and woods to premium markets that already trusted the Acacia label. It turned a once-abstract taxonomic vote into a potent commercial strategy, granting Australia's wattles privileged access to a global consumer base conditioned to equate Acacia with quality, reliability and versatility.

The door has already been opened and it's only a matter of time before the scammers walk in.

References:

1. CBI – Ministry of Foreign Affairs. What Is the Market Potential for Acacia Gum on the European Market? Centre for the Promotion of Imports from Developing Countries, July 25, 2023.
2. Hay, Thomas O., et al. "A New Hydrocolloid to Rival Gum Arabic: Characterization of a Traditional Food Gum from Australian Acacia cambagei." Food Hydrocolloids 153 (2023).
3. Hurr, T. O. "Acacia pycnantha Gum Exudates Recognised as a Traditional Food." Qeios, 2023. https://doi.org/10.32388/O9FH1K.
4. Brand, J. C., et al. Prospects for Commercial Production of Australian Native Plant Gums. Rural Industries Research and Development Corporation, 2013. RIRDC Publication No. 13-009.
5. Maiden, J. H. The Useful Native Plants of Australia: Including Tasmania. Sydney: Turner & Henderson, 1889. National Library of Australia digitized edition.

CHAPTER 18

EULOGY TO LESLIE PEDLEY
THE ACACIA HERO

Les Pedley was not merely a botanist, he was a visionary who stood at the nexus of science, ethics, and advocacy. He battled fiercely against overwhelming tides of interest groups. His fight against Australian interest groups intent on appropriating the name "Acacia" was not merely academic. It was a crusade grounded in foresight, integrity, and unwavering belief in historical truth. Pedley understood, long before others even sensed the threat of Proposal 1584, that nomenclature was more than taxonomy; it was heritage, identity, and legacy.

Before the Vienna Congress in 2005, when lobbyists wielded influence and persuasion with deft hands, Pedley alone emerged as a vocal guardian of authenticity. He confronted the shady tactics cloaked in legality but steeped in calculated maneuvering. He knew the game well and better, perhaps, than those who opposed him realized. Unlike many African and Asian botanists, tragically absent from a struggle that directly impacted their scientific legacy; Pedley stood resolute and called out what he recognized as subtle yet insidious lobbying.

His publications were not mere scholarly documents; they were robust interventions. When he famously reclassified Australian species into the genus Racosperma, it wasn't a simple scientific statement, but a bold and strategic act of resistance. Pedley was forthright, precise, and emphatic, arguing that "Acacia" ought to properly remain with the African thorn trees upon which Linnaeus first based the genus" [1]. His words echoed with conviction, illustrating a deep respect for historical and scientific precedence, declaring emphatically that the appropriation by Australian groups would jeopardize nomenclatural stability and blur the universality that the Code is meant to safeguard. He insisted that botanical names must not be bent to national sentiment but should be anchored in historical precedent.

Pedley foresaw clearly what many could not; the dangerous path paved by commercial interests that could lead to historical erasure. While lobbyists maneuvered behind closed doors, Pedley brought the fight openly into scholarly journals, elucidating clearly the ethical dimensions often hidden behind technical jargon. He insisted that botanical nomenclatures must reflect natural relationships, not political convenience. His efforts correlated with the majority of voters in the 2005 IBC congress. They all warned that to alter names for commercial and nationalistic interests would confuse science and erode the historical record.

Les Pedley's legacy transcends botanical taxonomy; it embodies integrity, resilience, and an unwavering commitment to truth. His foresight regarding the dangers posed by relinquishing historical names to commercial interests was prophetic. He must have recognized that the battle for the name "Acacia" was emblematic of a broader fight against historical revisionism driven by economic power and nationalistic ambitions. His heroism lay not just in his scholarship, but in his unyielding commitment to defend what was historically and ethically correct. Pedley exemplified the profound power of an individual standing against the tides of opportunism.

He leaves behind more than a scientific legacy. He leaves a lesson in bravery, advocacy, and the profound importance of integrity. In honoring Leslie Pedley, we recognize not merely a scientist, but a man whose voice challenged the powerful, whose pen illuminated truth, and whose heart preserved the rightful heritage of the African thorn tree, a name, a symbol, and a legacy: Acacia.

References:

1. Pedley, Les. "New combinations for Senegalia Raf. and Vachellia Wight & Arn. species (Mimosaceae) that occur in Australia." Austrobaileya, vol. 6, no. 3, 2003, pp. 484.

CHAPTER 19

THE POINT OF NO RETURN

There is a pattern to what civilizations choose to remember. Sometimes it takes the shape of a colossal statue shipped across oceans, or an ancient temple raised brick by brick to a new land. Other times, it is a word, a name, that carries the freight of memory. Few names have done more silent work across the span of human history than Acacia.

Not everyone who has spoken the name knew what they were invoking. They were always reaching for something sacred, something that spanned civilizations and creeds. The thorn tree of the Nile with its bark bleeding golden gum, became more than just a tree. It became the connective tissue binding civilizations. And yet, today, we are being asked to give the name away and sever the thread that ties the Nile valley civilizations to wandering prophets, Greek philosophers, Monastery scrolls, medieval Islamic scientists, and Enlightenment explorers. All in the name of botanical reclassification. However, this is not just science. This is memory and we've spent far more, moved far faster, to protect far less.

In 1968, Egypt dismantled the ancient temples of Abu Simbel stone by stone and reassembled them above rising Lake Nasser. UNESCO raised what would now be over $400 million in today's money to move the monuments 65 meters higher. Helicopters, Italian engineers, and German funding converged. All to preserve a symbol.

In Iraq, thousands of clay tablets and cylinder seals from Nineveh were secretly flown out of warzones and stored in London basements. In China, UNESCO funded the stabilization of the Mogao Caves at Dunhuang (a

1,600-year-old Buddhist manuscript archive) installing dehumidifiers and visitor limits to protect pigments laid by monks who died 1600 years before. These efforts were not about utility. They were about continuity.

Compare that to what is being asked now: to preserve a single word. One name. Acacia.

The Name begins in the Tree's origin, Ancient Egypt Who knew its value long before Moses. The Acacia to them was a vessel of divinity. Its heritage carried across millennia. In the Book of Exodus, when Yahweh commands Moses to construct the Ark of the Covenant, He specifies the wood to be used: "shittim wood" (Exodus 25:10). In Hebrew, it is שִׁטִּים (shittim), which scholars and archaeobotanists have long identified as the thorn tree of the Sinai and translated into countless books throughout Millennia as the "Acacia wood" of African/Asian Acacia. When the Hebrew Bible was translated into Greek (the Septuagint) around the 3rd century BCE it rendered the word as "ξύλων ἀκακίας" (xylon akakias), literally "wood of the Acacia". From that moment, the word Akakia entered the sacred vocabulary of both East and West.

The Latin Vulgate followed suit centuries later. The tree that once bled in Sinai became lignis acaciae, its sanctity solidified through translation. From Akakia the word eventually evolved into the English Acacia. However, before becoming Acacia it would first evolve into Aqaqiya in early Arabic pharmacology, and Acaciae in Enlightenment era Europe. This highlights a linguistic path that spread not by trade, but by reverence. The thorn tree is more than a plant. It is a linguistic artifact.

When Christian scribes preserved the names of healing plants in Syriac medical scrolls, they often transliterated Akakia directly. Muslim scholars retained it too, noting the golden resin in their pharmacopoeias. Trade documents from Alexandria and Venice, even into the 16th century, listed succus acaciae as a critical item on merchant ledgers.

In Masonic tradition, the Acacia is the symbol of immortality and resurrection. At the grave of Hiram Abiff (the legendary master builder of Solomon's Temple) it is an Acacia sprig that marks the burial. The tree becomes a symbol not just of life, but of righteous suffering, divine purpose, and eternal return. Masonic temples in New York and London still etch the Acacia into their insignia. Esoteric texts from the Rosicrucians and Hermeticists likewise anchor their secret teachings around the thorn tree of the east.

These are not fringe beliefs. They are part of the collective subconscious of civilizations who knew, instinctively, that Acacia was more than nomenclature. It was a revelation.

These names endure not because of classification, but because of function and faith. No Roman apothecary or bible translation wrote "Vachellia". They wrote Acacia because that was the tree that mattered. Now consider what it would mean to reassign that name to a species the European Union has now placed on its Invasive Species list. Trees that dry up aquifers and are being eradicated en masse to save the environment. Most tragically they carry no cultural weight. No mystic ever rested under their shade. No scroll ever praised their gum. No ancient "Gods" born within their trunk. Nothing!

Returning to the moment. The world has already spent billions preserving sandstone Buddhas, digitizing Sumerian tablets, and restoring the burnt fragments of Notre-Dame's roof. We protect because we remember. We remember because we name and names are the ligaments of our collective soul. To strip Acacia from its roots is to sever the sacred and to unwrite 2,000 years of collective history. Modern taxonomy was weaponized to invite confusion where there once was continuity. Unlike the operations saving the temples of Abu Simbel, we do not need helicopters, nor cranes. No Giant stones need be moved. The cost is infinitesimal. All it takes is the will to recognize that the name itself is a world and that some worlds are worth keeping.

All is not lost, and we can still do something to reclaim the Acacia legacy. I'm not a Botanist and I will not pretend to have a solution, but I do believe that if enough people feel strongly about anything then all is possible. The only thing we can do at this point is to make it known that we believe the name should be reclaimed. No amount of Botanical nomenclature entries will exceed the amount of work needed to revise and reinterpret all the religious books, historical books, trade manuals, patent documents, regulatory databases, product SKUs, and ancient script translations from 4,000-year-old temples. There is simply no comparison.

The vision to restore Acacia will ensure that the linguistic artifact that united civilizations, continents and humanity for 2,300 years is not lost and that our children will continue to learn from this legacy instead of ending it and placing it in a Botanical trash bin. It will also ensure that the much-needed healing products associated with the thorn tree are never appropriated by false marketers. We simply don't have any other option.

All we can do is ensure the IBC and every single Botanist and Herbaria understand we are all behind them. 4,000+ years of documented thorn tree history and 2,300 years of Acacia name history deserve this effort. The name that united civilizations through linguistics, culture, medicine, religion and trade deserves this effort.

There is already a growing coalition of botanists that believe the simplest fix is to restore the name Acacia to the original thorny trees and supply new, automatic combinations for the Australian wattles. The next IBC in 2030 can make that change. However, this will only happen if delegates are ensured of strong, broad support.

What we ask of you:
1. **Add your name** by scanning the QR code below and then fill it out. Your signature is the only way to inform the IAPT, IBC, national botanical societies, and taxonomic committees that clarity and fairness matter.
2. **Become an advocate** by creating a video talking about this subject or posting pictures with the hashtag #True_Acacia. Make sure you tag us on social media (links available on www.TrueAcacia.com or email us so we can help you share it).
3. **Share the QR code and link** on social media with #TrueAcacia hashtag so colleagues, students, and conservation groups can sign too.
4. **Reach out** to local herbaria, universities, and botanical gardens. Ask them to endorse the initiative and forward it to their international networks.
5. **Share this book with Botanists you know** as it will be the botanists who ultimately decide the fate of this matter
6. **Stay informed** – we'll update signatories on key deadlines (public-comment periods, delegate briefings) so you can send supportive notes at the right time.

RESTORE THE NAME, SUPPORT THE INITIATIVE
#TRUEACACIA

Scan the Code now
Sign the Initiative
Together we will reclaim Acacia

CONTACT US & FOLLOW OUR SOCIAL MEDIA HERE
www.TrueAcacia.com

A PLEA TO AUSTRALIA

Australia has, in recent decades, shown a profound willingness to reckon with its past. Where once the law declared this continent "terra nullius," the same courts now affirm ancient ownership and custodianship. Across the country, land is being returned not as a political concession, but as an act of scientific and historical clarity. Indigenous communities once written out of maps are being written back in, not as a romantic gesture but as a rectification. It is not perfect, and it is not yet finished but in recognizing the deep continuity between people and place, Australia has begun leading the world in what true reconciliation might look like.

Land is not just property. It is memory, it is inheritance, it is truth made tangible. And names (especially ancient ones) are much the same. A genus name, once anchored to a species, becomes a container of meaning. It defines how the world organizes knowledge. It tells us where something began. It tells us, quite literally, what it is. When that name is reassigned or displaced after 2,300 years of usage, it is not a neutral shift. It is like land reassignment, a transfer of recognition and authority.

What if a name were land? What if the genus Acacia, long rooted in the dry savannas of the Sahel, in the gum-harvesting practices of Kordofan, in the pharmacopoeias of Egypt, Europe, and West Asia; had never been relocated to Australia's wattles, but instead remained where it began? What if the name had followed the same "land title" logic of belonging to where the oldest evidence says it began?

I write this not to revisit the history as that has already been examined in this book, and it need not be reargued here. I write this final chapter instead as an appeal to the botanical community. Especially to those who have given their lives to Acacia, who have nurtured Australia's wattles through decades of taxonomic revision and ecological research. Bruce Maslin and his colleagues did not steal anything. Their drive to preserve the name Acacia for the largest group of species was from their nationalistic point of view. No one questions the integrity, or the labor involved in the work they have done.

However, perhaps now, after twenty years, space has opened to revisit that decision through another lens. A lens not of numerical convenience, but one of historical and ethical reconciliation. Australia has already shown that institutions can grow. The same nation that once justified dispossession under law now returns sacred lands with ceremony. The same parks that in the past excluded traditional Indigenous owners now include them as co-managers. What we thought was fixed turned out to be changeable. What we thought was irreversible proved to be correctable.

Returning the name Acacia to its original African lineage is not a repudiation of Australian science. It is an extension of its maturity. It is a recognition that taxonomy, like any knowledge system, does not exist in a vacuum. It reflects the values and assumptions of its time. And if we are bold enough to admit when those assumptions were incomplete, then we elevate taxonomy itself. We strengthen its credibility. We prove that science is not only self-correcting, but also self-aware.

Twenty years have passed since the International Botanical Congress redefined the genus Acacia, yet the world has already cast its vote, and it did not follow that decision. Regulatory bodies, scientific research, product labeling, and billions of global consumers continue to associate the name Acacia with its original African species. Despite the nomenclatural shift, Acacia gum is still Acacia gum in every pharmacopoeia, customs document, food ingredient list, and trade ledger. Acacia fiber remains on health supplement shelves with no nod to the name change. Research papers, even those published in the world's leading scientific journals, overwhelmingly continue to cite the African species under their historical names.

In effect, the world has refused to conform to the IBC's 2005 decision, revealing a dissonance that persists between botanical taxonomy and global economic, scientific, and regulatory realities. What purpose does this dissonance serve? Why should taxonomy stand isolated in defiance of consensus? Instead of trying to bend the world to fit into an unpopular scientific decision, it is time for botany to realign itself with the consensus that has already emerged. This is not a stubborn refusal of change, it is a recognition that the change imposed was, and remains, out of step with the living, breathing reality of global usage. Until this mistake is reversed, the confusion will continue not just for consumers, but for science itself.

No one is asking that Australia's wattles be diminished. They are already celebrated scientifically, ecologically, and culturally. Their diversity is astonishing. Their value is undisputed. However, they can flourish under scientifically coherent names like Racosperma, names that reflect the biogeographic distinctions now well established. To return Acacia to Africa is not to erase Australia's contribution. It is to decenter Australia's convenience in favor of global equity. It is, in short, to do what the nation has already begun doing with land.

This is not a demand. It is a gesture. It is a call to conscience, to continuity, and to quiet leadership. The legacy of Acacia is older than any modern taxonomy. Its story reaches back through millennia of medicinal use, trade, religious symbolism, and economic function. The African Acacias were the original custodians of this name. It was theirs long before Linnaeus put it to paper, and long before the 2005 vote reclassified it.

Australia has shown the world how to return what was wrongly taken. It has done so with humility, with care, and with remarkable foresight. The same can be done here. It will not be without cost, but no act of restoration ever is. What it offers, however, is something far rarer, the chance to make history whole. And in so doing, to set a new global standard not just for taxonomy, but for taxonomic justice.

Let the name go home. Not because it was stolen maliciously, but because now we know better and we can reconcile by leading with integrity. Reconciliation, whether of land or name, is only real when it is complete.

Throughout this book, I have passionately argued against the appropriation of the term "Acacia," driven by a deep commitment of preserving the historical and cultural legacy of the thorn tree that has shaped civilizations for millennia. My critique has targeted specifically the lobbying actions and the decision makers whose actions, I believe, risk obscuring an invaluable part of our shared human heritage.

To my Australian readers and friends, my intent was never to diminish the genuine love and pride you rightfully hold for your extraordinary native flora. Australia's landscapes and biodiversity, rich and unique, deserve global recognition and celebration. In using the term "Australians," I sought only shorthand for a specific group of lobbyists, not an indictment of an entire nation or its people.

I deeply respect Australia's commitment to preserving its natural heritage, and my aim remains the same; to protect and honor the integrity of cultural and botanical history worldwide. I extend my heartfelt appreciation and goodwill to all Australians who, like many around the globe, strive to uphold truth, history, and conservation in harmony.

ACKNOWLEDGMENTS

This book would not have been possible without the great publications by Gerry Moore, Bruce Maslin, Gideon Smith, Melissa Luckow, Anthony Orchard, Jane Carruthers, Pieter Winter, Renee Fortunato, Christopher William, Johan Hurter, Christian Kull, Haripriya Rangan, Gideon Smith, and most important of them all is Leslie Pedley, may God rest his soul. Not all these great people will agree with my assessments. However, we are all part of this story. I hope all will also be a part of this initiative to reclaim Acacia. Thank you all.

ABOUT THE AUTHOR

Mohamad Alnoor is a True Acacia enthusiast. He is also a consultant in IT systems, Real estate, Import & Export, Marketing, Logistics and E-commerce.

www.MohamadAlnoor.com

Support me by Tipping 😊 Thanks in Advance

https://paypal.me/RealMohamadAlnoor

DISCLAIMER

The author and publishers of The Acacia Heist have made every effort to ensure that the information presented herein is accurate, complete, and current as of the date of publication. Nevertheless:

Informational Purpose Only
The Work is provided solely for educational, historical, and journalistic purposes. Nothing contained herein should be construed as legal, financial, medical, or other professional advice. Readers should consult qualified professionals before acting on any information.

No Warranty of Accuracy or Completeness
All facts, figures, and citations are presented "as is" without warranties of any kind (express or implied) including but not limited to warranties of accuracy, reliability, completeness, merchantability, or fitness for a particular purpose.

Evolving Science & Policy
Scientific classifications, regulatory frameworks, and legal standards referenced in the text may change over time. The author and publisher accept no responsibility for updates after the publication date.

Good-Faith Scholarship & Fair Use
Quotations, images, tables, and other materials reproduced from third-party sources are used under fair-use, academic, or license-based principles. All trademarks, service marks, and trade names are the property of their respective owners; use herein does not imply endorsement.

Opinions & Interpretations
Analyses, viewpoints, and conclusions expressed are those of the author and do not necessarily reflect the views of any affiliated institution, sponsor, or reviewer. Any resemblance to real persons, organizations, or events beyond documented facts is coincidental.

Limitation of Liability
To the maximum extent permitted by law, the author, publisher, editors, and distributors disclaim all liability for any loss, damage, or disruption (direct or indirect) that may arise from using, relying on, or acting upon information contained in the Work.

Notice of Potential Errors
Despite rigorous sourcing, errors or omissions may remain. Readers who identify possible inaccuracies are invited to contact the publisher so corrections can be considered for future editions.

Copyright © 2025 by Mohamad Alnoor. All rights reserved. No part of this publication may be reproduced, stored in a retrieval system, or transmitted in any form or by any means.

www.ingramcontent.com/pod-product-compliance
Lightning Source LLC
Chambersburg PA
CBHW050522100526
44581CB00002B/78